2

QUADERNI

MONOGRAPHS

Umberto Zannier
Scuola Normale Superiore
Piazza dei Cavalieri, 7
56126 Pisa, Italia

Clemens Fuchs
University of Salzburg
Department of Mathematics
Hellbrunnerstr. 34/I
5020 Salzburg, Austria

On Some Applications of Diophantine Approximations

On Some Applications of Diophantine Approximations

(a translation of Carl Ludwig Siegel's *Über einige Anwendungen diophantischer Approximationen* by Clemens Fuchs)

edited by
Umberto Zannier

with a commentary and the article *Integral points on curves: Siegel's theorem after Siegel's proof* by Clemens Fuchs and Umberto Zannier

EDIZIONI
DELLA
NORMALE

ISBN: 978-88-7642-519-6
e-ISBN 978-88-7642-520-2

Contents

Preface

In 1929 Carl Ludwig Siegel published the paper *Über einige Anwendungen diophantischer Approximationen* (appeared in Abh. Preuß. Akad. Wissen. Phys.-math. Klasse, 1929 and reproduced in Siegel's collected papers Ges. Abh. Bd. I, Springer-Verlag 1966, 209-266). It was devoted to Diophantine Approximation and applications of it, and became a landmark work, also concerning a number of related subjects. Siegel's paper was written in German and this volume is devoted to a translation of it into English (including also the original version in German).

Siegel's paper introduced simultaneously many new methods and ideas. To comment on this in some detail would occupy a further paper (or several papers), and here we just limit ourselves to a brief discussion. The paper is, roughly, divided into two parts:

(a) The first part was devoted to proving transcendence of numbers obtained as values (at algebraic points) of certain special functions (including hypergeometric functions and Bessel functions). In this realm, Siegel's paper systematically developed ideas introduced originally by Hermite in dealing with the exponential function; a main point is to approximate with rational functions a function expressed by a power series, and to draw by specialisation numerical approximations to its values. These numerical approximations, if accurate enough, allow, through standard comparison estimates, to prove the irrationality (or transcendence) of the values of the function in question.

Siegel exploited and extended this principle in great depth, obtaining results which were (and are) spectacularly general in the topic, especially at that time.

Probably this was the first paper giving to transcendence theory some coherence.

This study also introduced related concepts, like the one of 'E-function' and of 'G-function' (*cf.* [1,8]), and led naturally to algebraical and arithmetical studies on systems of linear differential equations (with polynomial coefficients), which eventually inspired several different directions of research, all important and deep.

(b) The second part was devoted to diophantine equations, more precisely to the study of integral points o algebraic curves. In this realm too, the paper introduced several further new ideas, also with respect to previous important papers of Siegel.[1] Some instances for the new ideas contained in the second part are:

- The paper used the embedding of a curve (of positive genus) into its Jacobian, and the finite generation of the rational points in this last variety (which had been proved by L. J. Mordell for elliptic curves and by A. Weil in general); this provided a basic instance of the intimate connection of diophantine analysis with algebraic geometry and complex analysis, which became unavoidable since that time.
- It used and developed the concept of 'height' of algebraic points and its properties related to rational transformations, especially on algebraic curves; this went sometimes beyond results by A. Weil, who had introduced the concept and had also pointed out transformation properties.
  Again, this represented one of the very first examples of how the link between arithmetic and geometry can be used most efficiently, leading to profound results.
- It exploited to a new extent the diophantine approximation to algebraic numbers. This had been used by A. Thue around 1909 in the context of the special curves defined by the so-called 'Thue equations'. For general curves, the diophantine approximation drawn from integer solutions is not sharp enough to provide directly the sought information, hence Siegel had to go much deeper into this. By taking covers of the curve (inside its Jacobian) he improved the Diophantine approximation; then he was able to conclude through a suitable refinement of Thue's theorem, also proved in the paper. This refinement was not as strong as K.F. Roth's theorem (proved only in 1955), creating further complications to the proof; Siegel had to use simultaneous approximations to several numbers to get a sharp enough estimate. (For this task, Siegel used ideas introduced in the first part of the paper.)

On combining all of this, Siegel produced his theorem on integral points on curves, which may be seen as a final result in this direction. At that time, this was especially impressive, since Diophantine equations were often treated by *ad hoc* methods, with little possibility of embracing whole families. (One of the few exceptions occurred with Thue's methods, alluded to above.)

The theorem also bears a marked geometrical content: *an affine curve may have infinitely many integral points only if it has non-negative Euler characteristic*. (This is defined as $2 - 2g - s$ where g is the genus of a smooth projective model of the curve and where s is the number of points at infinity, namely those

[1] For instance, Siegel had written (chronologically) right before another remarkable paper on integral points by proving their finiteness for hyperelliptic equations $y^2 = f(x)$ under appropriate assumptions (this result was extracted from a letter to Mordell and was published under the pseudonym X in [47]).

missing on the affine curve with respect to a smooth complete model.) Conversely, this condition becomes sufficient (for the existence of infinitely many integral points) if we allow a sufficiently large number field and a sufficiently large (fixed) denominator for the coordinates.[2]

Each of the numerous results and ideas that we have mentioned would have represented at that time a major advance in itself. Hence it is difficult to overestimate the importance of this paper and its influence, even thinking of contemporary mathematics.

The paper, however, being written in German, is not accessible for a direct reading to all mathematicians. There are of course modern good or excellent expositions in English of some of the results, however we believe it may be of interest for many to go through the original source for a precise understanding of some principles, as really conceived by Siegel. In addition, we think that this paper is a model also from the viewpoint of exposition; the ideas and methods are presented in a limpid and simultaneously precise way.

All of this led the second author to the idea of translating the paper into English, and of publishing the result by the 'Edizioni della Normale'. After some partial attempts for a translation, this was finally carried out by the first author, and here is the output. The translation was made literally and keeping the style used by Siegel as far as possible, instead of rephrasing the text in more modern style. (In particular, Siegel is not using the engaging 'we' but instead uses the comprehensive 'one'. The reader should have this in mind when reading the translation.)

It has been eventually possible also to include the original text in this volume, which provides additional information.

We have also added to the translated text a small number of footnotes (marked as "FOOTNOTE BY THE EDITORS") to highlight a few points that we think are worth noting for the convenience of the reader. Further, after the translation, we have included into this publication an article by the two of us, describing some developments in the topic of integral points, and three modern proofs of Siegel's theorem (two of them being versions of the original argument). We have also inserted a few references to other work arising from the paper of Siegel.

ACKNOWLEDGEMENTS. We thank the Edizioni della Normale for having welcome this project. We especially thank Mrs. Luisa Ferrini of the 'Centro Edizioni', whose great care and attention have made it possible to reach the sought goal.

Clemens Fuchs and Umberto Zannier

[2]Note that *affine* implies $s > 0$; however, in view of Faltings' theorem - see 2.1 below - all of this remains true also for projective curves, *i.e.* when $s = 0$.

On some applications of Diophantine approximations*

Clemens Fuchs

Essays of the Prussian Academy of Sciences.
Physical-Mathematical Class 1929, No. 1

The well-known simple deduction rule according to which for any distribution of more than n objects to n drawers at least one drawer contains at least two objects, gives rise to a generalization of the Euclidean algorithm, which by investigations due to DIRICHLET, HERMITE and MINKOWSKI turned out to be the source of important arithmetic laws. In particular it implies a statement on how precisely the number 0 can be at least approximated by a linear combination

$$L = h_0\omega_0 + \cdots + h_r\omega_r$$

of suitable rational integers $h_0, \ldots, h_r$, which in absolute value are bounded by a given natural number H and do not vanish simultaneously, and of given numbers $\omega_0, \ldots, \omega_r$; in fact for the best approximation it certainly holds

$$|L| \leq (|\omega_0| + \cdots + |\omega_r|)H^{-r},$$

an assertion that does not depend on deeper arithmetic properties of the numbers $\omega_0, \ldots, \omega_r$.

The expression L is called an approximation form. If one then asks, how precisely the number 0 can at best be approximated by the approximation form $h_0\omega_0 + \cdots + h_r\omega_r$, then obviously any non-trivial answer will certainly depend on the arithmetic properties of the given numbers $\omega_0, \ldots, \omega_r$.

This question particularly contains the problem to investigate whether a given number ω is transcendental or not; one just has to choose $\omega_0 = 1, \omega_1 = \omega, \ldots, \omega_r = \omega^r, H = 1, 2, 3, \ldots, r = 1, 2, 3, \ldots$. The additional assumption, to come up even with a non-zero lower bound for $|L|$ as a function in H and r, gives a positive turnaround to the transcendence problem.

Also, the upper bound for the number of lattice points on an algebraic curve, thus in particular the study of the finiteness of this number, leads, as will turn out later, to the determination of a positive lower bound for the absolute value of a certain approximation form.

Analogous to the arithmetic problems of bounding $|L|$ from above and from below is an algebraic question. Let $\omega_0(x), \ldots, \omega_r(x)$ be series in powers of a

*CARL LUDWIG SIEGEL, *Über einige Anwendungen diophantischer Approximationen*, In: "Gesammelte Abhandlungen", Band I, Springer-Verlag, Berlin-Heidelberg-New York, 1966, 209–266.

variable x and let $h_0(x), \ldots, h_r(x)$ be polynomials of degree at most H, not all identically zero and having the property that the power series expansion of the approximation form

$$L(x) = h_0(x)\omega_0(x) + \cdots + h_r(x)\omega_r(x)$$

starts with a fairly large power of x; the goal is to get a lower and an upper bound for the exponent of this power of x. The algebraic problem is of easier nature than the arithmetic one; it leads to the determination of the rank of a system of linear equations.

These two problems, the algebraic and the arithmetic one, are connected by choosing for x a special rational number ξ from the common region of convergence of the power series and by assuming that the coefficients of these power series are rational numbers. Then in turn also the coefficients of the polynomials $h_0(x), \ldots, h_r(x)$ can chosen to be rational; and on multiplying with the common denominator of the rational numbers $h_0(\xi), \ldots, h_r(\xi)$ the algebraic approximation form $L(x)$ turns into an arithmetic one, unless the numbers $h_0(\xi), \ldots, h_r(\xi)$ are all equal to zero. However, in general the best algebraic approximation will not be turned into the best arithmetic approximation in this way.

To bound the expression $|h_0\omega_0 + \cdots + h_r\omega_r|$ from below under the conditions $|h_0| \leq H, \ldots, |h_r| \leq H$ there is the following possibility:

Let the numbers $\omega_0, \ldots, \omega_r$ be not all equal to zero. The $r+1$ approximation forms

$$L_k = h_{k0}\omega_0 + \cdots + h_{kr}\omega_r, \qquad (k = 0, \ldots, r)$$

shall be considered with coefficients that are rational integers and bounded in absolute value by H. Let the value of the determinant $|h_{kl}|$ be different from zero and denote by M the maximum of the $r+1$ numbers $|L_k|$. Let L be a further approximation form and let h be the largest among the absolute values of its coefficients. Since the $r+1$ forms $L_0, \ldots, L_r$ are linearly independent, one can choose r of them, say $L_1, \ldots, L_r$, which are linearly independent with L. Let (λ_{kl}) be the inverse of the matrix of the coefficients of $L, L_1, \ldots, L_r$; then the following estimates for the absolute value of the elements λ_{kl} hold

$$|\lambda_{k0}| \leq r!H^r \qquad (k = 0, \ldots, r)$$
$$|\lambda_{kl}| \leq r!hH^{r-1}. \qquad (k = 0, \ldots, r; l = 1, \ldots, r)$$

From the equalities

$$(1) \qquad \omega_k = \lambda_{k0}L + \lambda_{k1}L_1 + \cdots + \lambda_{kr}L_r, \qquad (k = 0, \ldots, r)$$

it follows that

$$(2) \qquad |L| \geq \frac{|\omega_k|}{r!H^r} - \frac{rMh}{H}.$$

If now for growing H the number M, which depends on H, is approaching 0 *faster than H^{1-r}, then* (2) *gives a positive lower bound for* $|L|$. This condition is therefore sufficient for the linear independence of the quantities $\omega_0, \dots, \omega_r$ over the field of rational numbers.

An analogous criterion holds for the linear independence of power series over the field of rational functions. In fact, let $\omega_0(x), \dots, \omega_r(x)$ be power series, not all identically zero; let

$$L_k(x) = h_{k0}(x)\omega_0(x) + \cdots + h_{kr}(x)\omega_r(x), \qquad (k = 0, \dots, r)$$

be $r + 1$ approximation forms with polynomial coefficients $h_{kl}(x)$ of degree H and let M be the smallest exponent which really appears in the power series expansions of $L_0, \dots, L_r$. Assume that the determinant $|h_{kl}(x)|$ is not identically zero. Let $L(x)$ be a further approximation form with coefficients of degree h. Let μ and μ_k be the smallest exponents in the power series $L(x)$ and $\omega_k(x), k = 0, \dots, r$. On observing that the determinant $|h_{kl}|(x)$ has degree $rH + h$, then the estimate

$$rH + h + \mu_k \geq \min(\mu, M) \qquad (k = 0, \dots, r) \tag{3}$$

follows from an equality similar to (1).

If now the difference $M - rH$ diverges for growing H, then (3) gives an upper bound for μ. In particular, this is sufficient for the linear independence of the power series $\omega_0(x), \dots, \omega_r(x)$ over the field of rational functions.

When applying this criterion the difficulty lies in the claim of the non-vanishing of the determinant $|h_{kl}(x)|$. To get to cases in which this difficulty can be mastered, from now on the functions $\frac{d\omega_o(x)}{dx}, \dots, \frac{d\omega_r(x)}{dx}$ are assumed to be expressible homogeneously and linearly by the functions $\omega_0(x), \dots, \omega_r(x)$ themselves and moreover with coefficients that are rational functions in x. It then holds a homogeneous system of first order linear differential equations

$$\frac{d\omega_k}{dx} = a_{k0}\omega_0 + \cdots + a_{kr}\omega_r; \qquad (k = 0, \dots, r) \tag{4}$$

and by differentiating an approximation form $L(x)$ another one is obtained, if one multiplies by the polynomial that appears as common denominator of the coefficients of $\omega_0, \dots, \omega_r$. Iterating this r-times, one has in sum $r + 1$ approximation forms. However, it can happen that the determinant of this system of $r + 1$ approximation forms is identically zero; then also the determinant $\Delta(x)$ of the system of the $r + 1$ linear forms $L, \frac{dL}{dx}, \dots, \frac{d^rL}{dx^r}$ vanishes and vice versa. The importance of identical vanishing of Δ follows from the following *lemma*:

Let

$$\omega_k = c_0\omega_{k0} + \cdots + c_r\omega_{kr}, \qquad (k = 0, \dots, r)$$

where $c_0, \dots, c_r$ are arbitrary constants, be the general solution of the system (4). *The determinant of the system of the $r + 1$ linear forms $L(x) =$*

$h_0(x)\omega_0(x)+\cdots+h_r(x)\omega_r(x)$, $\frac{dL}{dx},\ldots,\frac{d^rL}{dx^r}$ of $\omega_0,\ldots,\omega_r$ vanishes identically if and only if the $r+1$ functions

$$f_l = h_0\omega_{0l}+\cdots+h_r\omega_{rl} \qquad (l=0,\ldots,r)$$

are related by a homogeneous linear equation with constant coefficients.

If

$$\frac{d^kL}{dx^k} = b_{k0}\omega_0+\cdots+b_{kr}\omega_r, \qquad (k=0,\ldots,r)$$

with $b_{0l}=h_l$, $b_{k+1,l}=\frac{db_{kl}}{dx}+b_{k0}a_{0l}+\cdots+b_{kr}a_{rl}$ for $k=0,\ldots,r-1$ and $l=0,\ldots,r$, then from the assumption of $|b_{kl}|=\Delta$ vanishing identically one gets an equation

$$g_0\frac{d^sL}{dx^s}+g_1\frac{d^{s-1}L}{dx^{s-1}}+\cdots+g_sL=0,$$

where $s\le r$ and $g_0,\ldots,g_s$ denote certain sub-determinants of $\Delta(x)$ from which g_0 does not vanish identically. The function

$$L=\sum_{k=0}^r h_k\omega_k=\sum_{k=0}^r h_k\sum_{l=0}^r c_l\omega_{kl}=\sum_{l=0}^r c_lf_l,$$

satisfies this linear differential equation of order s, so do any of the $r+1$ functions $f_0,\ldots,f_r$; since their number is bigger than s, they are related by a homogeneous linear equation with constant coefficients. Conversely, it follows from such a relation by differentiating r-times, that the determinant $\left|\frac{d^kf_l}{dx^k}\right|$ is identically zero, and from the matrix relation

$$\left(\frac{d^kf_l}{dx^k}\right)=(b_{kl})(\omega_{kl})$$

one obtains the equation $|b_{kl}|=\Delta=0$, if one takes into account that the values of the solutions $\omega_{0l},\ldots,\omega_{rl}, l=0,\ldots,r$ in a regular point can be chosen so that the determinant $|\omega_{kl}|$ in this point is not 0 and hence is not identically zero.

An example is given by taking $\omega_k=e^{kx}$, so $\omega_{kl}=e^{kx}e_{kl}$, where (e_{kl}) denotes the identity matrix. Then $f_l=h_l(x)e^{lx}$ and there is no homogeneous linear equation with constant coefficients relating $f_0,\ldots,f_r$, since e^x is not an algebraic function. Thus the determinant Δ does not vanish identically in this case.

Now suppose again that the power series $\omega_0(x),\ldots,\omega_r(x)$ have only rational coefficients and that ξ is a rational number. Then one obtains from the system of the algebraic approximation forms for the functions $\omega_0(x),\ldots,\omega_r(x)$ a system of arithmetic approximation forms for the numbers $\omega_0(\xi),\ldots,\omega_r(\xi)$. The most important point of all these investigations is now the construction of approximation forms for which for the number $\Delta(\xi) \neq 0$ holds. For this one uses the following consideration, which was already used in a special case by THUE.

Let γ be the smallest exponent in the power series expansion of the approximation form L. Multiplying by the common denominator $N(x)$ of the rational functions a_{kl} from (4) one obtains from $\frac{dL}{dx}$ an approximation form L_1, whose power series does not contain powers in x smaller than $\gamma-1$. One now considers the determinant $D(x)$ of the $r+1$ approximation forms L, $N\frac{dL}{dx} = L_1$, $N\frac{dL_1}{dx} = L_2, \dots, N\frac{dL_{r-1}}{dx} = L_r$. If ν is an upper bound for the degree of the $(r+1)^2+1$ polynomials N, Na_{kl}, and if the coefficients of the form L have degree H, then $D(x)$ has degree $H + (H+\nu) + \cdots + (H + r\nu) = (r+1)H + \frac{r(r+1)}{2}\nu$. On the other hand, it is possible to express $D(x)\omega_k(x)$ linearly homogeneously in $L, L_1, \dots, L_r$, say

$$D\omega_k = \Lambda_{k0}L + \Lambda_{k1}L_1 + \cdots + \Lambda_{kr}L_r, \qquad (k = 0, \dots, r) \tag{5}$$

and moreover with coefficients Λ_{kl}, which are polynomials in x. Therefore the function $D(x)\omega_k(x)$ vanishes at $x = 0$ with order at least $\gamma - r$. If one supposes that not all the power series $\omega_0, \dots, \omega_r$ are divisible by x, which otherwise could be eliminated by division, then it follows that $D(x)$ vanishes at a non-zero value $x = \xi$ with order at most $(r+1)H + \frac{r(r+1)}{2}\nu + r - \gamma$, unless $D(x)$ is identically equal to zero. Now, through a suitable choice of L, it is possible to obtain $\gamma \geq (r+1)h + r$; in fact the $r+1$ polynomials $h_0(x), \dots, h_r(x)$ of degree H contain $(r+1)(H+1)$ numbers as coefficients, which necessarily satisfy γ homogeneous linear equations. But then $D(x)$ vanishes at $x = \xi$ with order s, which is below the H-independent bound $\frac{r(r+1)}{2}\nu$, and the sth derivative of $D(x)$ is not zero at $x = \xi$. Moreover, equation (5) holds *identically* in $\omega_0, \dots, \omega_r$; differentiating it s-times and using (4) to eliminate the derivatives of $\omega_0, \dots, \omega_r$, then the resulting equation also holds identically in $\omega_0, \dots, \omega_r$. Putting also $N\frac{dL_r}{dx} = L_{r+1}, \dots, \frac{dL_{r+s-1}}{dx} = L_{r+s}$, then by (5) the expression $N^s(\xi)D^{(s)}(\xi)\omega_k$ ia a homogeneous linear relation of $L(\xi), L_1(\xi), \dots, L_{r+s}(\xi)$ identically in $\omega_0, \dots, \omega_r$. Assuming further that $N(\xi) \neq 0$, *i.e.* that ξ is different from the singular points of the system (4), then one obtains $\omega_0, \dots, \omega_r$ as linear relation of $L(\xi), L_1(\xi), \dots, L_{r+s}(\xi)$; among the $r+s+1$ forms $L(\xi), \dots, L_{r+s}(\xi)$ there are hence $r+1$ which are linearly independent. In this way one finds $r+1$ arithmetic approximation forms for the numbers $\omega_0(\xi), \dots, \omega_r(\xi)$ with determinant $\neq 0$.

For applications in number theory it is still necessary that the approximation forms, constructed above, lead to favorable arithmetic approximations in the sense explained earlier, *i.e.* that the coefficients of $L(x), \dots, L_{r+s}(x)$ do not contain "too large" rational integer numbers. Since the number s is below a bound which does not depend on H, essentially one just needs a good estimate for the coefficients of the polynomials $h_0(x), \dots, h_r(x)$ in $L = h_0\omega_0 + \cdots + h_r\omega_r$.

This can be easily demonstrated in the previously mentioned example $\omega_k(x) = e^{kx}$, because the coefficients can be expressed explicitly in terms of r and H; in this way one obtains a proof of the transcendence of e in the first of HERMITE's versions and, at the same time, a positive lower bound for the distance of an arbitrary algebraic number to e. It should also be noted that from HERMITE's formulae one immediately gets the transcendence of π and even a positive lower bound for the distance of an arbitrary algebraic number to π, when one takes into account that the norm of a non-zero algebraic integer has absolute value ≥ 1.

Another example, but only in the case $r = 1$, is given by the well-known continued fractions for the quotients of hypergeometric series. In particular, the continued fraction expansion of the function $(1 - x)^\alpha$ was used by THUE to investigate the approximation of roots of natural numbers by rational numbers, and this was the starting point for the discovery of THUE's theorem on Diophantine equations.

In other cases one does not find an estimate beyond the trivial one for the numerical coefficients of the algebraically favorable approximation form, whose power series is divisible by $x^{(r+1)(H+1)-1}$, and the trivial bound is not sufficient, as one easily sees, in order to apply the arithmetic criterion. Therefore the strategy, that has led to a system of arithmetic approximation forms for $\omega_0(\xi), \dots, \omega_r(\xi)$ having non-vanishing determinant, has to be slightly modified. One has to optimize between the two necessities of a good algebraic and arithmetic approximation to 0 by admitting for the number γ a smaller value than the one taken earlier, for the number s thus a bigger one, and so one gets better bounds for the coefficients of $h_0(x), \dots, h_r(x)$ as a gain. This idea again was first applied by THUE. The estimate for the coefficients is obtained by using DIRICHLET's deduction rule, which was mentioned at the beginning, and which is here demonstrated in the form of a *lemma*:

Let

$$\begin{aligned} y_1 &= a_{11}x_1 + \cdots + a_{1n}x_n \\ &\vdots \\ y_m &= a_{m1}x_1 + \cdots + a_{mn}x_n \end{aligned}$$

be m linear forms in n variables with rational integer coefficients. Let $n > m$. Let the absolute values of the mn coefficients a_{kl} be not bigger than a given natural number A. Then the homogeneous linear equations $y_1 = 0, \dots, y_m = 0$ are solvable in rational integer numbers $x_1, \dots, x_n$, which are not all zero, but are all smaller than $1 + (nA)^{\frac{m}{n-m}}$ in absolute value.

For the proof let each of the variables $x_1, \dots, x_n$ independently run through the values $0, \pm 1, \dots, \pm H$; one obtains in sum $(2H + 1)^n$ lattice points in the space given by orthogonal Cartesian coordinates $y_1, \dots, y_m$, which however are not necessarily all different from each other. Each coordinate of each of these lattice points lies between the values $-nAH$ and $+nAH$. There are exactly $(2nAH + 1)^m$ different lattice points in the m-dimensional space, whose coordinate lie between $-nAH$ and $+nAH$. If now

$$(2nAH + 1)^m < (2H + 1)^n, \tag{6}$$

then two lattice points $y_1, \ldots, y_m$ belonging to different systems $x_1, \ldots, x_n$ coincide; and by subtracting these two systems one obtains a solution of $y_1 = 0, \ldots, y_m = 0$ in rational integer numbers $x_1, \ldots, x_n$, which are not all zero and in absolute value are $\leq 2H$. But the condition (6) is satisfied if the even integer number, which lies in the interval

$$(nA)^{\frac{m}{n-m}} - 1 \leq 2H < (nA)^{\frac{m}{n-m}} + 1,$$

is chosen for $2H$.

The method, which was sketched before, to determine a positive lower bound for the expression $|h_0\omega_0 + \cdots + h_r\omega_r|$ shall be applied in this exposition to two different problems. The first part mainly deals with the proof of the transcendence of the values of the cylindrical function evaluated at any non-zero algebraic number. The second part is concerned with the task of finding all algebraic curves which pass through infinitely many lattice points of the plane or, more generally, of the n-dimensional space; it will be shown that this can only happen in case of lines and hyperbolas and for certain other curves obtained from these by easy transformations and having also genus 0.

The motivation to study the problems in the first part came from W. MAIER's beautiful investigations on irrationality. The second part has its origin in the important results on the arithmetic properties of algebraic curves, which were discovered and recently published by A. WEIL in his thesis.

Part I: On transcendental numbers.

Dedicated to MAX DEHN.

By the theorems due to HERMITE and LINDEMANN the question of the arithmetic properties of the values of the exponential function at algebraic arguments has been answered. While the additivity of the exponential function reduces every algebraic equation between values of this function to a linear equation, something comparable does not exist anymore for other functions; and there lies the difficulty of generalizing HERMITE's arguments. For none of the other functions, which are of importance in calculus, a theorem of analogous strength like that for the exponential function has been found.

The irrationality of the cylindrical function

$$J_0(x) = \sum_{n=0}^{\infty} \frac{(-1)^n}{n!n!} \left(\frac{x}{2}\right)^{2n}$$

has been studied by different authors. HURWITZ and STRINSBERG proved that $J_0(x)$ is irrational for every non-zero rational value of x^2 and MAIER, going beyond that, showed in an extremely clever way that for such x the value $J_0(x)$ is not even a quadratic irrationality.

In what follows it will be shown that $J_0(x)$ is transcendental for every algebraic non-zero x. The method even gives the more general result that between the numbers $J_0(x)$ and $J_0'(x)$ no algebraic relation with rational coefficients can exist since it is shown that a positive lower bound for the absolute value of an arbitrary polynomial in $J_0(x)$ and $J_0'(x)$, whose coefficients are rational numbers, exists in terms of these coefficients. More generally, an analog of LINDEMANN's theorem will be shown, that between the numbers $J_0(\xi_1)$, $J_0'(\xi_1), \dots, J_0(\xi_k)$, $J_0'(\xi_k)$ no algebraic relation with rational coefficients exists if $\xi_1^2, \dots, \xi_k^2$ are pairwise distinct non-zero algebraic numbers.

The proof is done with the method that was discussed in the introduction. In particular to apply the first lemma a theorem is needed assuring that the function $J_0(x)$ is not solution of any first-order differential equation whose coefficients are polynomials in x. The previously stated theorem, that for algebraic non-zero values x no algebraic relation with algebraic coefficients exists between the numbers $J_0(x)$ $J_0'(x)$ and x, is merely a consequence of the theorem that assures that no algebraic equation identically in x between the functions $J_0(x)$ $J_0'(x)$ holds. This might indicate that even in more general cases numerical relations can be obtained by specialization of functional equations, so that calculus contains arithmetic in this sense.

§1. Tools from complex analysis.

In this paragraph it will be investigated which algebraic functional equations exist between the solutions of the BESSEL differential equation

$$\frac{d^2y}{dx^2} + \frac{1}{x}\frac{dy}{dx} + \left(1 - \frac{\lambda^2}{x^2}\right) y = 0, \tag{7}$$

their derivatives and the independent variable x. It will be shown that except the well-known relations no others exist, or - better said - that every other relation is obtained from these through a rational transformation.

Theorem 1. *Every algebraic function y satisfying the* BESSEL *differential equation is identically equal to* 0.

Proof. In the neighborhood of $x = \infty$ an expansion

(8) $$y = a_0 x^r + a_1 x^s + \cdots$$

with decreasing powers of x having integer or fractional exponents holds. Substituting (8) into (7) it follows that $a_0 = 0$, *i.e.* y has to vanish identically. □

Theorem 2. *Let y be a solution of the* BESSEL *differential equation* (7). *The functions y, $\frac{dy}{dx}$, and x satisfy an algebraic equation with constant coefficients if and only if λ is half of an odd number.*

Proof. Suppose that the equation

(9) $$P(y', y, x) = 0$$

holds, where P is an irreducible polynomial in the three unknowns y', y, x. By (7) y also satisfies the differential equation

(10) $$\frac{\partial P}{\partial y'}\left\{-\frac{1}{x}y' - \left(1 - \frac{\lambda^2}{x^2}\right)y\right\} + \frac{\partial P}{\partial y}y' + \frac{\partial P}{\partial x} = 0.$$

This again is an algebraic equation in y', y and x. By Theorem 1 it is not possible to eliminate y' from the equations (9) and (10). After multiplication by x^2 the left-hand side of (10) is a polynomial in y', y, x and hence must be divisible by the irreducible polynomial P. Considering its degree it follows at once that the quotient is a quadratic polynomial $ax^2 + bx + c$ just in x. The left-hand side of (10) equals $\frac{dP}{dx}$, whenever (7) is satisfied, hence for *every* solution of BESSEL's differential equation. Since (7) is homogeneous in y, y', y'', the equation

$$\frac{dP}{dx} = \left(a + \frac{b}{x} + \frac{c}{x^2}\right)P$$

is still satisfied for every solution of (7), when therein the polynomial P is replaced by the aggregation Q of the terms of highest degree in y', y in P. By integration it follows that

(11) $$Q(y', y, x) = kx^b e^{ax - \frac{c}{x}}$$

with k constant. Let now y_1 and y_2 be any two linearly independent solutions of BESSEL's differential equation; then $y = \lambda_1 y_1 + \lambda_2 y_2$ for arbitrary constant values λ_1 and λ_2 satisfies the equation (11). Since Q is a homogeneous non-constant polynomial in y', y, the integration-constant k in (11) is a homogeneous polynomial in λ_1 and λ_2. Let the ratio $\lambda_1 : \lambda_2$ be chosen so that $k = 0$. For the corresponding function y the following equation then holds

$$Q(y', y, x) = 0,$$

and consequently the quotient $y' : y$ is an algebraic function in x.

So if there is a solution of BESSEL's differential equation, which satisfies an algebraic differential equation of first order, then there is also such a solution, whose logarithmic derivative is an algebraic function. This theorem, by the way, is proved exactly in the same way for solutions of arbitrary homogeneous linear differential equation of second order with algebraic coefficients.

The logarithmic derivative $z = y' : y$ of every solution of BESSEL's differential equation satisfies RICCATI's equation

$$\frac{dz}{dx} + z^2 + \frac{1}{x}z + 1 - \frac{\lambda^2}{x^2} = 0. \tag{12}$$

If now z is algebraic, then at $x = \infty$ an expansion

$$z = a_0 x^{r_0} + a_1 x^{r_1} + \cdots \qquad (r_0 > r_1 > \cdots)$$

with decreasing powers of x having integer or fractional exponents and non-zero coefficients $a_0, a_1, \ldots$ holds.

It follows that

$$\sum_{k=0}^{\infty} a_k (r_k + 1) x^{r_k - 1} + \sum_{k=0}^{\infty}\sum_{l=0}^{\infty} a_k a_l x^{r_k + r_l} + 1 - \frac{\lambda^2}{x^2} = 0.$$

Comparison of coefficients first gives $r_0 = 0$, $a_0^2 + 1 = 0$, $a_0 = \pm i$; moreover for $n = 1, 2, \ldots$ the exponent r_n of the term $a_0 a_n x^{r_0 + r_n}$ has to be equal to one of the exponents $-2, r_k - 1, r_k + r_l$ $(k = 0, \ldots, n-1; l = 0, \ldots, n-1)$; and this implies that all the exponents $r_0, r_1, \ldots$ are integers. In particular $r_1 = -1$, $2a_0 a_1 + a_0 = 0$, $a_1 = -\frac{1}{2}$.

BESSEL's differential equation has singular points only at 0 and ∞; consequently also the algebraic function z can only be ramified in 0 and ∞. It has just been proved that every branch of z at infinity is regular. Hence, z is a rational function in x. Let

$$z = bx^s + \cdots$$

be the expansion at $x = 0$; then (12) gives the equation

$$b(s+1)x^{s-1} + b^2 x^{2s} - \frac{\lambda^2}{x^2} + \cdots = 0,$$

hence $s = -1$ and $b = \pm\lambda$. Since by (7) every root of y, not 0 or ∞, is of first order, z has in these zeros of y poles of first order with residue 1. Consequently,

$$z = \pm i \pm \frac{\lambda}{x} + \frac{1}{x - x_1} + \cdots + \frac{1}{x - x_k}$$

for certain non-zero constants $x_1, \ldots, x_k$. The expansion at $x = \infty$ gives $a_1 = \pm\lambda + k$, thus

$$\lambda = \pm\left(k + \frac{1}{2}\right)$$

with k a non-negative rational integer.

In order to have a relation between y', y, x holding identically in x, it is therefore necessary that λ is half of an odd number. But this is also sufficient, because it is well known that the linearly independent functions

$$H_1 = (-1)^k \frac{(2x)^{k+\frac{1}{2}}}{\sqrt{\pi}} \frac{d^k}{d(x^2)^k} \frac{e^{ix}}{ix}$$

and

$$H_2 = (-1)^k \frac{(2x)^{k+\frac{1}{2}}}{\sqrt{\pi}} \frac{d^k}{d(x^2)^k} \frac{e^{-ix}}{-ix}$$

satisfy BESSEL's differential equation with $\lambda = \pm(k + \frac{1}{2})$, and every homogeneous linear combination of H_1 and H_2 obviously has the form

$$y = \sqrt{x}(R(x)\cos x + S(x)\sin x),$$

where $R(x)$ and $S(x)$ are rational functions in x, such that y satisfies a first order differential equation

$$P_1(x)y'^2 + P_2(x)y'y + P_3(x)y^2 = P_4(x),$$

whose coefficients $P_1(x), \ldots, P_4(x)$ are polynomials in x. □

Theorem 3. *Suppose that λ is not half of an odd number and let y_1, y_2 be two linearly independent solutions of* BESSEL*'s differential equation* (7). *Then the functions y_1, $\frac{dy_1}{dx}$, y_2, x do not satisfy an algebraic equation with constant coefficients.*[1]

[1] FOOTNOTE BY THE EDITORS: Some ideas of differential algebra are explained in the proof in a very clear way. For more on this see e.g. [A. Buium, Differential algebra and Diophantine geometry, Current Mathematical Topics, Hermann, Paris, 1994].

Proof. Suppose that an equation

$$P(y_1, y_1', y_2, x) = 0 \tag{13}$$

holds, where P is a polynomial in the four unknowns y_1, y_1', y_2, x. By Theorem 2 the polynomial P indeed contains y_2. Let P be irreducible and of degree exactly $n \geq 1$ in y_2. Denote the coefficient of y_2^n by $f(y_1, y_1', x)$. From (13) and by differentiating w.r.t. x, it follows that

$$\frac{dP}{dx} = \frac{df}{dx} y_2^n + nf y_2^{n-1} y_2' + \cdots = 0.$$

But now with a constant $\alpha \neq 0$ it holds

$$y_1 y_2' - y_2 y_1' = \frac{\alpha}{x} \tag{14}$$

and therefore

$$\frac{dP}{dx} = \left(\frac{df}{dx} + nf\frac{y_1'}{y_1}\right) y_2^n + \cdots = 0. \tag{15}$$

Eliminating from this equation y_1'' through (7), one gets an algebraic equation involving y_1, y_1', y_2, x, which is also of degree n in y_2. However by Theorem 2, y_2 cannot be eliminated neither from this equation nor from (13). Consequently, the irreducible polynomial P differs from the function $\frac{dP}{dx}$, from which y_2' and y_1'' were eliminated by using (7) and (14), only by a factor free from y_2, namely by (15) evidently by the factor $f : \left(\frac{df}{dx} + nf\frac{y_1'}{y_1}\right)$. Therefore

$$\frac{P'}{P} = \frac{f'}{f} + n\frac{y_1'}{y_1}$$

holds *identically* in y_1, y_1', y_2, x whenever (7) and (14) are satisfied. One therefore can substitute y_2 by $y_2 + \lambda y_1$ in this equation, for arbitrary constant λ, and the identity remains valid. Integration gives

$$P(y_1, y_1', y_2 + \lambda y_1, x) = c(\lambda) f(y_1, y_1', x) y_1^n,$$

where $c(\lambda)$ denotes a polynomial in λ with constant coefficients. By differentiating w.r.t. λ and putting $\lambda = 0$, it follows that

$$y_1(nf y_2^{n-1} + \cdots) = c_1 f y_1^n$$

with c_1 constant. This is an equation of exact degree $n - 1$ in y_2. Since any equation involving y_1, y_1', y_2, x and truly containing y_2 has degree at least n in y_2, it follows that $n = 1$.

Hence, one has

(16) $$y_2 = g(y_1, y_1', x) : f(y_1, y_1', x),$$

where g and f are polynomials in y_1, y_1', x. Substituting this into (14) and replacing y_1'' by y_1' and y_1 with the help of BESSEL's differential equation, by Theorem 2 one obtains an identical equation in y_1', y_1, x. In particular the terms of highest degree in y_1 and y_1' in (14) must vanish. Assume that the rational function $g : f$ has degree d in y_1, y_1'. Because of the homogeneity of (7), the derivative of $g : f$ has also the same degree. Keeping in $g : f$ only the terms of highest degree both in the numerator and in the denominator, y_2 becomes a homogeneous rational function z in y_1, y_1' whose degree is d, and then the degree of $y_2 - z$ is smaller than d. The right-hand side of (14) has degree zero, consequently $d + 1 \geq 0$. If $d + 1 > 0$, then by (14)

$$y_1 z' - y_1' z = 0$$

holds; if however $d + 1 = 0$, then

$$y_1 z' - y_1' z = \frac{\alpha}{x}.$$

In the first case one has $z = by_1$ with b a constant, so $d = 1$; then one substitutes in (16) the function y_2 by the function $y_2 - by_1$ and is then reduced to the second case. In that case z is a solution of BESSEL's differential equation, which is linearly independent from y_1. Therefore there is a rational function in y_1, y_1', x, which is homogeneous of degree -1 in y_1, y_1' and satisfies equation (7). Denote it by $R(y_1, y_1', x)$. By Theorem 2 also $R(\lambda_1 y_1 + \lambda_2 y_2, \lambda_1 y_1' + \lambda_2 y_2', x)$ satisfies BESSEL's differential equation, hence one also has

(17) $$R(\lambda_1 y_1 + \lambda_2 y_2, \lambda_1 y_1' + \lambda_2 y_2', x) = \Lambda_1 y_1 + \Lambda_2 y_2,$$

where Λ_1 and Λ_2 only depend on λ_1 and λ_2. From the formulae

$$Ry_2' - \frac{dR}{dx} y_2 = \frac{\alpha}{x} \Lambda_1,$$
$$Ry_1' - \frac{dR}{dx} y_1 = -\frac{\alpha}{x} \Lambda_2$$

one recognizes that Λ_1 and Λ_2 are homogeneous rational functions in λ_1, λ_2 of degree -1. Hence, the ratio $\lambda_1 : \lambda_2$ can be chosen so that at least one of the functions Λ_1, Λ_2 becomes infinite. Since y_1 and y_2 are not proportional, also the right-hand side of (17) becomes infinite. With the values of λ_1 and λ_2 the equation

(18) $$I : R(y, y', x) = 0$$

holds for $y = \lambda_1 y_1 + \lambda_2 y_2$, and this contradicts Theorem 2. Hence, Theorem 3 is proved. □

The method used to prove Theorems 1, 2, 3 is put together from ideas of LIOUVILLE and RIEMANN. One can proceed analogously in case of an arbitrary homogeneous linear differential equation of second order with algebraic coefficients in order to find all algebraic relations involving the solutions, their derivatives and the independent variables.

The elementary algebraic idea, which led to the proof of Theorem 3, is maybe not completely clear. One can obtain this theorem in an essentially different way, which requires deeper arithmetic tools, but in its essence is quite easy to understand. In fact, it turns out that no algebraic relation involving y_1, y_1', y_2, x can exist, because the coefficients in the power series of y_2 are "too different" from the ones that appear in the power series of y_1. The proof of this fact is related to investigations done by EISENSTEIN and TSCHEBYSCHEFF; its idea is due to MAIER. First, as an obvious generalization of a theorem by EISENSTEIN, a sufficient condition for the algebraic independence of power series shall be established:[2]

Let

$$f(x) = \sum_{n=0}^{\infty} \gamma_n x^n$$

$$f_\nu(x) = \sum_{n=0}^{\infty} \gamma_n^{(\nu)} x^n \qquad (\nu = 1, \ldots, h)$$

be power series, whose coefficients all belong to a given algebraic number field $\mathfrak{K}$. If to any natural r there is an $n > r$ such that the exact denominator of γ_n contains a prime ideal from $\mathfrak{K}$, which does not divide the denominators of the numbers $\gamma_0, \ldots, \gamma_{n-1}$ and $\gamma_0^{(\nu)}, \ldots, \gamma_{n+r}^{(\nu)}$ $(\nu = 1, \ldots, h)$, then there is no algebraic relation, properly containing f, between $f, f_1, \ldots, f_h$.

For the proof suppose that there is an algebraic equation $P = 0$ involving $f, f_1, \ldots, f_h$ that contains f. Let its degree l be smallest possible with respect to f so that then $\frac{\partial P}{\partial f}$ does not identically vanish in x. Put

$$P = p_0 f^l + \cdots + p_l,$$

then the coefficients of the polynomials $p_0, \ldots, p_l$, which depend on $f_1, \ldots, f_h$, can be determined by homogeneous linear equations, which one obtains by substituting for $f, f_1, \ldots, f_h$ in the relation $P = 0$ the corresponding power series and by comparing coefficients afterwards. Therefore the coefficients of $p_0, \ldots, p_l$ can be chosen as integers of $\mathfrak{K}$. Let now

$$\frac{\partial P}{\partial f} = c_r x^r + \cdots$$

[2] FOOTNOTE BY THE EDITORS: We find it worth pointing out that a generalization of Eisenstein's theorem on denominators of coefficients of algebraic power series is obtained here.

with $c_r \neq 0$ be the power series of $\frac{\partial P}{\partial f}$. Put

$$\sum_{n=0}^{r} \gamma_n x^n = g,$$

then by TAYLOR's theorem one gets

$$P(f, ...) = P(g, ...) + (f-g)\frac{\partial P(g, ...)}{\partial g} + \frac{(f-g)^2}{2}\frac{\partial^2 P(g, ...)}{\partial g^2} + \cdots = 0,$$

$$(19) \qquad (f-g)\frac{\partial P(g)}{\partial g} = -P(g) - \frac{(f-g)^2}{2}\frac{\partial^2 P(g)}{\partial g^2} + \cdots.$$

Since the first $r+1$ terms in the series for f and g coincide, one also has

$$\frac{\partial P}{\partial g} = c_r x^r + \cdots.$$

Compare now the coefficients of x^{n+r} in (19), where n is a natural number $> r$. Obviously $c_r \gamma_n$ becomes a polynomial in $\gamma_0, \ldots, \gamma_{n-1}$ and $\gamma_0^{(\nu)}, \ldots, \gamma_{n+r}^{(\nu)}$ $(\nu = 1, \ldots, h)$ with integer coefficients from $\mathfrak{K}$. By assumption n can be chosen so that the exact denominator of γ_n contains a prime ideal $\mathfrak{p}_n$, which does not divide the denominators of the numbers $\gamma_0, \ldots, \gamma_{n-1}$ and $\gamma_0^{(\nu)}, \ldots, \gamma_{n+r}^{(\nu)}$. But then the denominator of $c_r \gamma_n$ contains the prime ideal $\mathfrak{p}_n$ neither, and hence the numerator of c_r is divisible by $\mathfrak{p}_n$. Since this holds true for infinitely many n and since the prime ideals $\mathfrak{p}_n$ corresponding to different n are different, one derives at a contradiction.

The second proof of Theorem 3 can be done as follows: Let $P(y_1, y_1', y_2, x) = 0$ be an algebraic equation involving y_1, y_1', y_2, x and truly containing y_2. If y_3, y_4 are two arbitrary linearly independent solutions of BESSEL's differential equation, then for constants p, q, r, s

$$y_1 = py_3 + qy_4, \qquad y_2 = ry_3 + sy_4$$

and by (14)

$$y_1' = py_3' + q\frac{y_4 y_3'}{y_3} + \frac{\alpha q}{ps - qr}\frac{1}{xy_3};$$

hence $P = 0$ is transformed into an algebraic equation in y_3, y_3', y_4, x, and conversely every equation in y_3, y_3', y_4, x can be written as equation in y_1, y_1', y_2, x. By Theorem 2 it is therefore sufficient, to consider for y_1 and y_2 two particular, non-proportional, solutions of (7).

If λ is a rational integer, then there is a solution y_1 that is regular and a solution y_2 that is *logarithmically* ramified at $x = 0$. This contradicts the fact that any solution y of the equation $P(y_1, y_1', y_2, x) = 0$ has at $x = 0$ the character of an algebraic function.

Therefore one can restrict to the case that 2λ is not an integer. Put

$$K_\lambda(x) = \sum_{n=0}^{\infty} \frac{(-1)^n}{n!(\lambda+1)(\lambda+2)\cdots(\lambda+n)} \left(\frac{x}{2}\right)^{2n}, \tag{20}$$

then

$$\frac{1}{\Gamma(\lambda+1)} \left(\frac{x}{2}\right)^{\lambda} K_\lambda(x) = J_\lambda(x)$$

is BESSEL's function, and J_λ and $J_{-\lambda}$ are two linearly independent solutions of (7). From an algebraic equation in $J_\lambda, J_\lambda', J_{-\lambda}, x$ containing $J_{-\lambda}$, an algebraic equation in $K_\lambda, K_\lambda', K_{-\lambda}, x$ follows, hence, since K_λ and $K_{-\lambda}$ are regular at $x = 0$, an algebraic equation

$$R(K_\lambda, K_\lambda', K_{-\lambda}, x) = 0,$$

now containing $K_{-\lambda}$. The coefficients $\xi_1, \ldots, \xi_q$ of the polynomial R are determined by comparing coefficients, hence from infinitely many homogeneous linear equations

$$\alpha_{k1}\xi_1 + \cdots + \alpha_{kq}\xi_q = 0, \qquad (k = 0, 1, \ldots) \tag{21}$$

in which the quantities α_{kl} $(k = 0, 1, \ldots; l = 1, \ldots, q)$ are rational integers with rational integer coefficients obtained from the coefficients of K_λ and $K_{-\lambda}$ and that are therefore rational functions in λ with rational coefficients. If all the determinants of the $q \times q$-submatrices of the matrix (α_{kl}) were identically zero in λ, then $\xi_1, \ldots, \xi_q$ could be chosen as polynomials in λ and so the equation $R = 0$ would hold identically in λ. If there is a $q \times q$-submatrix whose determinant does not vanish identically, then this is a rational function in λ with rational coefficients; but since the system (21) is solvable, this function has to be 0 and hence λ satisfies an algebraic equation with rational coefficients. In both cases one can therefore assume that λ belongs to an algebraic number field $\mathfrak{K}$, and then also the coefficients of K_λ and $K_{-\lambda}$ are numbers belonging to $\mathfrak{K}$.

Assume first that λ is rational, *i.e.* $\lambda = \frac{a}{b}$, $(a, b) = 1, b \geq 3$. By DIRICHLET's theorem there are infinitely many prime numbers of the form $bn - a$. Let $(b-1)n - a - 1 > 0$. For $m = 1, 2, \ldots, (b-1)n - a - 1$ one has $0 < m + n < p$; thus p does not divide $b(n+m) = p + bm + a$, thus neither $bm + a$ nor the denominators of the coefficients of $x^0, x^1, \ldots, x^{2(b-1)n-2a-2}$ in K_λ and $K_{-\lambda}$ by (20) for $p > 2$. Moreover, p is a factor of the denominator of the coefficient of x^{2n} in $K_{-\lambda}$, while the denominators of the previous coefficients are coprime to p. The lemma proved earlier implies that $K_{-\lambda}$ is not algebraically dependent of K_λ, K_λ', x.

Now assume that λ is an algebraic irrational and that

$$f(x) = a_0(x - \lambda_1)\cdots(x - \lambda_s) = a_0x^s + \cdots = 0$$

is an irreducible equation for λ with rational integer coefficients. By a theorem proved by NAGELL the largest prime divisor of the product $f(1)f(2)\cdots f(n)$ goes faster to infinity than n itself. In particular, one can find arbitrarily large natural numbers n such that $f(n)$ contains a prime p which is bigger than $3n$ and which does not divide $f(1),\dots,f(n-1)$. Moreover, assume that $p>|a_0|$ holds. Let $\mathfrak{p}$ be a prime ideal of $\mathfrak{K}$, which divides both p and $a_0(n-\lambda)$. For $m=1,2,\dots,2n$ one has $0<n+m\le 3n$, so that p is not a factor of $m+n$ and therefore neither of $a_0(n+m)=a_0(m+\lambda)+a_0(n-\lambda)$. Thus the prime ideal $\mathfrak{p}$ does not divide $a_0(m+\lambda)$. It follows that $\mathfrak{p}$ is a factor of the denominator of the coefficient of x^{2n} in $K_{-\lambda}$, but $\mathfrak{p}$ does not divide the denominators of the coefficients of $x^0,\dots,x^{2n-1}$ in $K_{-\lambda}$ and of $x^0,\dots,x^{4n}$ in K_λ and K'_λ. By the lemma $K_{-\lambda}$ is algebraically independent of K_λ, K'_λ, x also in this case.

This proves Theorem 3 a second time. The special case of the algebraic independence of J_λ, $J_{-\lambda}$, x was obtained even without appealing to Theorem 2.

Theorem 3 allows to apply the first lemma of the introduction. Let $l_1,\dots,l_\nu$ be ν different non-negative integers and, moreover, let $f_{kl}(x)$ be $(l_1+1)+\cdots+(l_\nu+1)=q$ polynomials, where the indices range over the values $k=0,\dots,l$ and $l=l_1,\dots,l_\nu$. Suppose that for no l among $l_1,\dots,l_\nu$ the $l+1$ polynomials $f_{0l},\dots,f_{ll}$ are all identically zero. With a solution y of BESSEL's differential equation now put

$$\phi_\rho = f_{0l}y'^{l} + f_{1l}y'^{l-1}y + \cdots + f_{ll}y^{l} \tag{22}$$

with $l=l_\rho$ and $\rho=1,\dots,\nu$ and then sum

$$\phi = \phi_1+\phi_2+\cdots+\phi_\nu = P(y,y'). \tag{23}$$

This is a polynomial in y and y', which in fact only contains terms of degree $l_1,\dots,l_\nu$; its coefficients are the q polynomials f_{kl}. The q functions $w_{kl}=y^k y'^{l-k}$ $(k=0,\dots,l;\ l=l_1,\dots,l_\nu)$ satisfy a system of q homogeneous linear differential equations of first order; in fact by (7) one has

$$\frac{dw_{kl}}{dx} = kw_{k-1,l} - \frac{l-k}{x}w_{kl} - (l-k)\left(1-\frac{\lambda^2}{x^2}\right)w_{k+1,l}. \tag{24}$$

Since in each differential equation of this system, there are only functions with the same index l, the system splits up into ν completely independent single systems, and in such a subsystem only the functions $w_{0l},\dots,w_{ll}$ appear. Then equations (24) are satisfied if we substitute in $w_{kl}=y^k y'^{l-k}$ for y any solution of BESSEL's differential equation, *i.e.* $y=\lambda_1y_1+\lambda_2y_2$, where y_1 and y_2 are two linearly independent solutions and λ_1,λ_2 are arbitrary constants. Putting for the ease of notation

$$\sum_{\rho=0}^{l}\binom{k}{\rho}\binom{l-k}{r-\rho} y_1^{\rho}y_2^{k-\rho}y_1'^{r-\rho}y_2'^{l-k-r+\rho} = \psi_{krl}, \qquad (r=0,\dots,l) \tag{25}$$

then

$$
\begin{aligned}
w_{kl} &= (\lambda_1 y_1 + \lambda_2 y_2)^k (\lambda_1 y_1' + \lambda_2 y_2')^{l-k} \\
&= \sum_{r=0}^{l} \lambda_1^r \lambda_2^{l-r} \psi_{krl}, \qquad (k = 0, \ldots, l)
\end{aligned} \tag{26}
$$

and, since these functions satisfy identically in λ_1, λ_2 the equations (24), for any fixed l

$$
w_{kl} = \sum_{r=0}^{l} c_r \psi_{krl} \qquad (k = 0, \ldots, l)
$$

is a solution of (24), where $c_0, \ldots, c_l$ are some constants. But this is the general solution, because from

$$
\sum_{r=0}^{l} c_r \psi_{krl} = 0 \qquad (k = 0, \ldots, l)
$$

it follows particularly for $k = l$ that

$$
\sum_{r=0}^{l} c_r \binom{k}{r} y_1^r y_2^{k-r} = 0,
$$

hence by the linear independence of y_1, y_2 that

$$
c_0 = 0, \ldots, c_l = 0.
$$

In this way also the general solution of the complete system (24) has been found.

According to (22), (23) and (24) the functions $\frac{d^s \phi}{dx^s} = \phi^{(s)}$ for $s = 1, 2, \ldots$ are again polynomials in y, y' only containing terms of degree $l_1, \ldots, l_\nu$, therefore homogeneous linear forms of the w_{kl}. Now it will be proved that the determinant of the q forms $\phi, \phi', \ldots, \phi^{(q-1)}$ that are obtained for $s = 0, 1, \ldots, q-1$ is not identically zero. By the first lemma in the introduction it is enough to show the linear independence of the q functions obtained from

$$
\phi = \sum_{k,l} f_{kl} w_{kl}
$$

by substituting for $l = l_\rho$ the functions $w_{0l}, \ldots, w_{ll}$ by $\psi_{0rl}, \ldots, \psi_{lrl}$, for $l \neq l_\rho$ by 0, and by choosing $\rho = 1, \ldots, \nu$ and $r = 0, \ldots, l$.

Assume the validity of the equation

$$
\sum_{l} \sum_{r=0}^{l} \sum_{k=0}^{l} c_{rl} f_{kl} \psi_{krl} = 0 \tag{27}
$$

with c_{rl} constants, then one obtains from it an algebraic equation involving y_1, y_1', y_2, x by eliminating through (14) the function y_2'. By Theorem 3 one gets an identity in the four variables y_1, y_1', y_2, x. In it only the terms of highest degree in y_1, y_1', y_2 shall be considered. By (14) one gets those by keeping only the largest l in (27), for which the constants $c_{cl}, \ldots, c_{ll}$ are not all zero, and on substituting in the expression for ψ_{krl} given by (25) the function y_2' by $\frac{y_2}{y_1}y'$. Considering (26), it follows that

$$\sum_{r=0}^{l}\sum_{k=0}^{l} c_{rl} f_{kl} \binom{l}{r} y_1^r y_2^{l-r} \left(\frac{y_1'}{y_1}\right)^{l-k} = 0$$

holds identically in $y_1, y_2, \frac{y_1'}{y_1}$. Therefore

$$c_{rl} f_{kl} = 0, \qquad (r = 0, \ldots, l; k = 0, \ldots l)$$

and this is a contradiction, since neither the constants $c_{0l}, \ldots, c_{ll}$, nor the polynomials $f_{0l}, \ldots, f_{ll}$ are all identically zero. Hence, one has:

Theorem 4. *Let λ be not half of an odd number and let y be a solution of* BESSEL*'s differential equation* (7). *Consider the expression*

$$\phi = \sum_{l}\sum_{k=0}^{l} f_{kl} y^k y'^{l-k},$$

in which the coefficients f_{kl} are polynomials in x, and contain only degrees $l = l_1, \ldots, l_\nu$ of y and y', so that ϕ is a homogeneous linear form[3] *of the $(l_1 + 1) + \cdots + (l_\nu + 1) = q$ monomials $y^k y'^{l-k}$ $(k = 0, \ldots, l; l = l_1, \ldots, l_\nu)$. Then also every derivative of ϕ is such a homogeneous linear form and the determinant of the q forms $\phi, \phi', \ldots, \phi^{(q-1)}$ is not identically zero.*

Obviously this theorem can also be derived for the solutions of other homogeneous linear differential equations of second order with algebraic coefficients.

§2. Tools from arithmetic.

In order to make the usage of the second lemma from the introduction possible, the coefficients of the functions $y^k y'^{l-k}$ will be studied from an arithmetic perspective. The tool to do this is given by the following theorem due to MAIER:

Theorem 5. *Let α and γ be rational numbers; let γ be different from $0, -1, -2, \ldots$. Let h_n be the least common denominator of the n fractions*

$$\frac{\alpha}{\gamma}, \frac{\alpha(\alpha+1)}{\gamma(\gamma+1)}, \ldots, \frac{\alpha(\alpha+1)\cdots(\alpha+n-1)}{\gamma(\gamma+1)\cdots(\gamma+n-1)}. \qquad (n = 1, 2, \ldots)$$

Then the growth of h_n is less than the nth power of a suitable constant.

[3] FOOTNOTE BY THE EDITORS: Actually, *homogeneous* here means that Φ is a sum of homogeneous terms of degrees $l_1, \ldots, l_\nu$ in y, y' respectively.

Proof. Let $\alpha = a : b, \gamma = c : d, (c, d) = 1, d > 0$. Because of the equation

$$\begin{aligned} &\frac{\alpha(\alpha+1)\cdots(\alpha+l-1)}{\gamma(\gamma+1)\cdots(\gamma+l-1)}\cdot\frac{b^{2l}}{d^l} \\ &\quad = \frac{a(a+b)\cdots(a+(l-1)b)b^l}{c(c+d)\cdots(c+(l-1)d)} \qquad (l = 1, 2, \ldots, n) \end{aligned} \tag{28}$$

it is sufficient to show the statement for the least common denominator of the right-hand side of (28). The denominator $c(c+d)\cdots(c+(l-1)d) = N_l$ is not divisible by d. Let p be a prime factor of N_l. As ν varies among any p^k consecutive rational integer numbers, then exactly one of the p^k numbers $c+\nu d$ is divisible by p^k. Of the l factors of the denominator N_l at least $[lp^{-k}]$ and at most $[lp^{-k}]+1$ will be divisible by p^k; however, for $p^k > |c|+(l-1)d$ none is divisible by p^k. For the exponent s of the power of p which divides N_l, the inequality

$$\sum_k [lp^{-k}] \le s \le \sum_k ([lp^{-k}]+1)$$

holds, wherein k varies through all natural numbers satisfying the condition $p^k \le |c|+(l-1)d$. Therefore for constants c_1 and c_2

$$\left[\frac{l}{p}\right] \le s < \left[\frac{l}{p}\right] + c_1\frac{l}{p^2} + c_2\frac{\log l}{\log p}.$$

The numerator $a(a+b)\cdots(a+(l-1)b)b^l = Z_l$ is divisible by p to an exponent which is at least $[l/p]$. Consequently, in the denominator of the reduced fraction $Z_l : N_l$ the exponent of p is smaller than $c_1\frac{l}{p^2} + c_2\frac{\log l}{\log p}$. One obtains for the logarithm of the least common denominator H_n of the n fractions $Z_1 : N_1, \ldots Z_n : N_n$ the estimate

$$\log H_n < \sum_p \left(c_1\frac{n}{p^2} + c_2\frac{\log n}{\log p}\right)\log p = c_1 n\sum_p \frac{\log p}{p^2} + c_2\log n\sum_p 1,$$

where p varies through all prime numbers below the bound $|c|+(n-1)d$. By an elementary theorem in the theory of prime numbers it holds

$$\sum_p 1 < c_3\frac{n}{\log n};$$

moreover, the sum $\sum \frac{\log p}{p^2} = c_4$ taken over all prime numbers is convergent. Therefore

$$\log H_n < c_1c_4 n + c_2c_3 n = (c_1c_4 + c_2c_3)n,$$

which completes the proof. □

Now consider power series

$$y = \sum_{n=0}^{\infty} \frac{a_n}{b_n} \frac{x^n}{n!}$$

with the following properties:

1. The numerators $a_0, a_1, \ldots$ are integers of a fixed algebraic number field and the absolute values of all the conjugates of the a_n grow with n slower than any fixed positive power of $n!$;
2. The denominators $b_0, b_1, \ldots$ are natural numbers and the least common multiple of $b_0, \ldots, b_n$ also grows with n slower than any fixed positive power of $n!$;
3. The function y satisfies a linear differential equation, whose coefficients are polynomials with algebraic coefficients.

A function y, whose power series has these three properties, shall briefly be called an E-function. Obviously the exponential function is an E-function. Every E-function is an entire function. The E-functions have several important, partly evident, properties that will be mentioned now; for this let $E(x)$ denote an arbitrary E-function.

I. Every algebraic constant is an E-function.
II. For algebraic constants α, $E(\alpha x)$ is an E-function.
III. The derivative $E'(x)$ is an E-function.
IV. The integral $\int_0^x E(t)dt$ is an E-function.
V. If $E_1(x)$ and $E_2(x)$ are E-functions, then $E_1(x)+E_2(x)$ is an E-function.
VI. If $E_1(x)$ and $E_2(x)$ are E-functions, then $E_1(x)E_2(x)$ is an E-function.

Of these statements only V. and VI. need a proof. Let

$$E_1(x) = \sum_{n=0}^{\infty} \frac{a'_n}{b'_n} \frac{x^n}{n!}, \quad E_2(x) = \sum_{n=0}^{\infty} \frac{a''_n}{b''_n} \frac{x^n}{n!}, \tag{29}$$

so by putting

$$a'_n b''_n + a''_n b'_n = a_n, \quad b'_n b''_n = b_n, \quad E_1(x) + E_2(x) = y \tag{30}$$

one has

$$y = \sum_{n=0}^{\infty} \frac{a_n}{b_n} \frac{x^n}{n!}.$$

Since a'_n, a''_n, b'_n, b''_n satisfy the conditions 1. and 2., also a_n satisfies condition 1. Denoting by $\{p, \ldots, q\}$ the least common multiple of the natural numbers $p, \ldots, q$, one has

$$\{b'_0 b''_0, \ldots, b'_n b''_n\} \leq \{b'_0, \ldots, b'_n\}\{b''_0, \ldots, b''_n\},$$

therefore b_n also satisfies condition 2. If the linear differential equations for E_1 and E_2 have order h_1 and h_2 respectively, then every derivative of y is a linear combination of $E_1, E_1', \ldots, E_1^{(h_1-1)}, E_2, E_2', \ldots, E_2^{(h_2-1)}$ with coefficients rational functions in x having algebraic coefficients. Consequently, y satisfies a linear differential equation of order h_1+h_2, whose coefficients are polynomials with algebraic coefficients. This proves V.

To prove VI. put in contrast to (30)

$$\{b_0', \ldots, b_n'\}\{b_0'', \ldots, b_n''\} = b_n \tag{31}$$

$$b_n \sum_{k=0}^{n} \binom{n}{k} \frac{a_k' a_{n-k}''}{b_k' b_{n-k}''} = a_n, \tag{32}$$

hence by (29)

$$y = E_1 E_2 = \sum_{m=0}^{\infty}\sum_{n=0}^{\infty} \binom{m+n}{n} \frac{a_m' a_n''}{b_m' b_n''} \frac{x^{m+n}}{(m+n)!} = \sum_{n=0}^{\infty} \frac{a_n}{b_n}\frac{x^n}{n!}.$$

It holds $\{b_0, \ldots, b_n\} = b_n$ and by (31) condition 2. is satisfied. Using $\sum_{k=0}^{n}\binom{n}{k} = 2^n$ and (32), condition 1. is satisfied. Finally, condition 3. is also satisfied, because every derivative of y can be linearly expressed through the $h_1h_2+h_1+h_2$ functions $E_1, E_1', \ldots, E_1^{(h_1-1)}, E_2, E_2', \ldots, E_2^{(h_2-1)}, E_1E_2, \ldots, E_1^{(h_1-1)}E_2^{(h_2-1)}$.

From I,. . ., VI. it follows

Theorem 6. *Let $E_1(x), \ldots, E_m(x)$ be any E-functions and let $\alpha_1, \ldots, \alpha_m$ be algebraic numbers. Then every polynomial with algebraic coefficients in $E_1(\alpha_1 x), \ldots, E_m(\alpha_m x)$ and the derivatives of these functions is again an E-function.*

Every polynomial in x with algebraic coefficients is in particular an E-function. One obtains a non-trivial example of an E-function by applying Theorem 5. Pick l rational numbers $\gamma_1, \ldots, \gamma_l$, different from 0, −1, −2, . . ., and k rational numbers $\alpha_1, \ldots, \alpha_k$. Let $l-k=t>0$. Put $c_n=0$, if n is not a multiple of t, and

$$c_n = \prod_{p=1}^{k}(\alpha_p(\alpha_p+1)\cdots(\alpha_p+m-1)) : \prod_{q=1}^{l}(\gamma_q(\gamma_q+1)\cdots(\gamma_q+m-1)),$$

if $n=mt$ is divisible by t. In order to see that

$$y = \sum_{n=0}^{\infty} c_n x^n \tag{33}$$

is an E-function, write c_n for $n = mt$ in the form

$$(34)\qquad \begin{aligned} c_n = {} & \frac{\alpha_1 \cdots (\alpha_1 + m - 1)}{\gamma_1 \cdots (\gamma_1 + m - 1)} \cdots \frac{\alpha_k \cdots (\alpha_k + m - 1)}{\gamma_k \cdots (\gamma_k + m - 1)} \cdot \\ & \cdot \frac{1 \cdots m}{\gamma_{k+1} \cdots (\gamma_{k+1} + m - 1)} \cdots \frac{1 \cdots m}{\gamma_l \cdots (\gamma_l + m - 1)} \cdot \frac{(mt)!}{(m!)^t} \cdot \frac{1}{n!}. \end{aligned}$$

Let a_n/b_n be the reduced fraction $n!c_n$, then by Theorem 5 a_n and b_n satisfy conditions 1. and 2. The fact that condition 3. is also satisfied follows from the form of the c_n, from which the linear differential equation for y can easily be obtained.

Hence, BESSEL's function

$$J_0(x) = \sum_{n=0}^{\infty} \frac{(-1)^n}{n!n!} \left(\frac{x}{2}\right)^{2n}$$

is an E-function and, more generally, the function

$$K_\lambda(x) = \Gamma(\lambda + 1) \left(\frac{x}{2}\right)^{-\lambda} J_\lambda(x) = \sum_{n=0}^{\infty} \frac{(-1)^n}{n!(\lambda + 1) \cdots (\lambda + n)} \left(\frac{x}{2}\right)^{2n}$$

for every rational value of λ, different from $-1, -2, \ldots$, is an E-function.

It would be interesting to see an E-functions which is not obtained from the special E-functions defined by (33) and (34) through the operations given by I,. . ., VI.

One may ask about the algebraic relations between given E-functions $E_1(x), \ldots, E_m(x)$, that is about the existence of polynomials in the variables $x_1, \ldots, x_m$ vanishing identically in x when $x_1, \ldots, x_m$ are replaced by the functions $E_1(x), \ldots, E_m(x)$. Such a relation is for instance

$$x^2 K_{\frac{1}{2}}^2 + K_{-\frac{1}{2}}^2 = 1.$$

By Theorem 3 a very special instance of this general problem has been settled, namely that an algebraic equation involving $K_\lambda, K_\lambda', K_{-\lambda}, x$ for $\lambda \neq 0, \pm 1, \pm 2, \ldots$ holds if and only if 2λ is an odd number. A deeper theorem about BESSEL functions will be given in the sequel.

It seems that the following question lies much deeper: Let $\alpha_1, \ldots, \alpha_m$ be algebraic numbers; the goal is to determine if there is an algebraic equation with rational coefficients involving the *numbers* $E_1(\alpha_1), \ldots, E_m(\alpha_m)$. This problem contains the one mentioned before because equations involving $E_1(x), \ldots, E_m(x)$, identically satisfied in x, can be written with algebraic coefficients and are in particular satisfied for algebraic x. One can formulate this also in a slightly different way. Every monomial in $E_1(\alpha_1 x), \ldots, E_m(\alpha_m x)$ is again an E-function; the problem is therefore transformed into the task to decide whether or not the values $E_1(1), \ldots, E_m(1)$ are linearly independent over the field of rational numbers. The treatment of this question is done using the method that has been outlined in the introduction, the applicability only requires an assertion of the type like the one in Theorem 4. Thereby, the arithmetical problem of a numerical equation is turned into the algebraic problem of a functional equation to be identically satisfied in x. As an example for this one can take the exponential function; according to LINDEMANN's theorem all algebraic equations $P(e^{\alpha_1}, \dots, e^{\alpha_m}) = 0$ are algebraic consequences of the functional equation $\exp(x + y) = \exp(x)\exp(y)$.

Approximation forms for E-functions shall now be given, which give a good approximation to 0 algebraically as well as arithmetically.

Theorem 7. *Let k E-functions $E_1(x), \dots, E_k(x)$ with rational coefficients be given. Let n be a natural number. There are k polynomials $P_1(x), \dots, P_k(x)$ of degree $2n - 1$ with the following properties:*

1. *The coefficients of $P_1(x), \dots, P_k(x)$ are rational integers, not all zero and at most of order $(n!)^{2+\varepsilon}$ as functions of n, where ε is a given, arbitrarily small, positive number;*
2. *one has*

$$P_1 E_1 + \cdots + P_k E_k = \sum_{\nu=(2k-1)n}^{\infty} q_\nu \frac{x^\nu}{\nu!}, \tag{35}$$

where the left-hand side E-function $P_1 E_1 + \cdots + P_k E_k$ vanishes at $x = 0$ with order at least $(2k - 1)n$;

3. *the coefficients q_ν have as functions of n and ν at most order $(n!)^2(\nu!)^\varepsilon$.*

Proof. Let

$$E(x) = \sum_{n=0}^{\infty} \gamma_n \frac{x^n}{n!}$$

be one of the k functions $E_1, \dots, E_k$. Let $g_0, \dots, g_{2n-1}$ be rational integers and put

$$P(x) = (2n - 1)! \sum_{\nu=0}^{2n-1} g_\nu \frac{x^\nu}{\nu!}, \tag{36}$$

$$d_l = \sum_{\rho=0}^{2n-1} \binom{l}{\rho} g_\rho \gamma_{l-\rho}; \tag{37}$$

then one has

$$P(x)E(x) = (2n - 1)! \sum_{l=0}^{\infty} d_l \frac{x^l}{l!}. \tag{38}$$

If the coefficients of $x^0, x^1, \dots, x^{(2k-1)n-1}$ in the power series for $P_1 E_1 + \cdots + P_k E_k$ are required to vanish, then $(2k - 1)n$ homogeneous linear equations for the $2kn$ unknown coefficients of the k polynomials of degree $2n - 1$

have to be satisfied. The least common denominator of the rational numbers $\gamma_0, \dots, \gamma_{(2k-1)n-1}$ is $O((n!)^\varepsilon)$; the same estimate holds for the binomial coefficients $\binom{l}{\rho}$ with $\rho = 0, \dots, l$ and $l = 0, \dots, (2k-1)n - 1$ and hence also for the integral coefficients of the $(2k-1)n$ homogeneous linear equations. By the second lemma in the introduction, these equations can be solved in terms of the rational integer values of the unknown variables g_ν, where not all are 0 and they have order

$$(2kn(n!)^\varepsilon)^{\frac{(2k-1)n}{2kn-(2k-1)n}}. \tag{39}$$

The polynomials so obtained have the three stated properties. By (36) and (39) 1. is satisfied; moreover, 2. is satisfied, and, since $\gamma_\nu = O((\nu!)^\varepsilon)$, from (37) and (38) the estimate stated in 3. follows for the coefficient q_ν of $\frac{x^\nu}{\nu!}$ on the right-hand side of (35). □

The meaning of Theorem 7 is that on the one hand expansion (35) starts with a *high* power of x and that on the other hand the rational integer coefficients of the polynomials $P_1, \dots, P_k$ are *small*. One could even obtain that the expansion (35) starts with the power x^{2kn-1}, but then the coefficients of $P_1, \dots, P_k$ would probably not be so small anymore, as claimed in Theorem 7 under the assumption 1.; and it is this aspect which will be crucial in the sequel.

Let l be a natural number and let $k = \frac{(l+1)(l+2)}{2}$. Denote by $E_1(x), \dots$ $\dots, E_k(x)$ the k functions

$$J_0^\kappa J_0'^{\lambda-\kappa}, \qquad (\kappa = 0, \dots, \lambda; \lambda = 0, \dots, l)$$

and apply Theorem 7 to them. Then for any n there are k, not all identically vanishing, polynomials $f_{\kappa\lambda}$ $(\kappa = 0, \dots, \lambda; \lambda = 0, \dots, l)$ of degree $2n-1$ with rational integer coefficients of order $(n!)^{2+\varepsilon}$, so that the power series for the function

$$\phi(x) = \sum_{\kappa,\lambda} f_{\kappa\lambda}(x) J_0^\kappa J_0'^{\lambda-\kappa} \tag{40}$$

starts with the power $x^{(2k-1)n}$ and is bounded from above by

$$O\left((n!)^2 \sum_{\nu=(2k-1)n} \frac{|x|^\nu}{(\nu!)^{1-\varepsilon}}\right). \tag{41}$$

Because of the differential equation $xJ_0'' = -J_0' - xJ_0$ the functions $x^a\phi^{(a)}(x)$ for $a = 1, 2, \dots$ are polynomials in x, J_0, J_0', namely in x of degree $2n + a - 1$, in J_0, J_0' of degree l. The coefficients of these polynomials are again rational integers and for $a < n + k^2$ of order $(n!)^{1+\varepsilon} \cdot (n!)^{2+\varepsilon}$. The power series for $x^a\phi^{(a)}(x)$ also starts with $x^{(2k-1)n}$ and is bounded from above by (41) times $n!$.

Denote by $t_1, \dots, t_k$ the k monomials $J_0^\kappa J_0'^{\lambda-\kappa}$ ordered in an arbitrary way; then $x^a\phi^{(a)}(x)$ is a homogeneous linear function in $t_1, \dots, t_k$, say

$$x^a\phi^{(a)}(x) = \sigma_{a1}(x)t_1 + \dots + \sigma_{ak}(x)t_k.$$

In the sequel, let

$$2n \geq k^2 \tag{42}$$

hold.

Theorem 8. *Let ξ be a non-zero number. Among the $n+k^2$ linear forms*

$$\sigma_{a1}(\xi)t_1 + \dots + \sigma_{ak}(\xi)t_k \qquad (a = 0, 1, \dots, n+k^2-1)$$

in the k variables $t_1, \dots, t_k$ there are k linearly independent forms.

Proof. One has to use Theorem 4 and to repeat the arguments used at the end of the introduction. In the function $\phi(x)$ defined by (40) let the variables J_0, J_0' only appear with the degrees $l_1, l_2, \dots, l_\nu$. It follows that only the $(l_1+1) + \dots + (l_\nu+1) = q$ monomials $J_0^\kappa J_0'^{\lambda-\kappa}$ $(\kappa = 0, \dots, \lambda;\ \lambda = l_1, l_2, \dots, l_\nu)$ appear, which will be denoted by $\omega_1, \dots, \omega_q$. Hence

$$x^a\phi^{(a)}(x) = \tau_{a1}(x)\omega_1 + \dots + \tau_{aq}(x)\omega_q, \tag{43}$$

where $\tau_{a1}, \dots, \tau_{aq}$ are polynomials in x of degree $2n+a-1$. The determinant

$$D(x) = |\tau_{ab}(x)|,$$

where the row-index takes the values $a = 0, \dots, q-1$, the column-index takes the values $b = 1, \dots, q$, has degree in x equal to

$$q(2n-1) + 1 + 2 + \dots + (q-1).$$

If $T_{ba}(x)$ denotes the determinant of $\tau_{ab}(x)$, then by (43)

$$D(x)\omega_b = T_{b0}(x)\phi(x) + T_{b1}(x)x\phi'(x) + \dots + T_{bq-1}(x)x^{q-1}\phi^{(q-1)}(x) \tag{44}$$

for $b = 1, \dots, q$. By Theorem 4 the determinant $D(x)$ is not identically 0. Moreover, $x^a\phi^{(a)}(x)$ vanishes at $x = 0$ at least of order $(2k-1)n$; therefore the same is true for the right-hand side of (44). If b is chosen such that ω_b is one of the powers $J_0^{l_1}, \dots, J_0^{l_\nu}$, then ω_b does not vanish at $x = 0$. Hence, $D(x)$ is divisible by $x^{(2k-1)n}$, and it holds that

$$q(2n-1) + 1 + 2 + \dots + (q-1) - (2k-1)n = \delta \geq 0. \tag{45}$$

For $x = \xi \neq 0$ let $D(x)$ vanish with order s; then it follows that $s \leq \delta$ and $D(\xi) = 0, \dots, D^{(s-1)}(\xi) = 0, \dots, D^{(s)}(\xi) \neq 0$.

Moreover, one has

$$q = (l_1 + 1) + \cdots + (l_\nu + 1) \leq 1 + \cdots + (l + 1) = k, \tag{46}$$

and equality holds only if the numbers $l_1, \ldots, l_\nu$ coincide with the numbers $0, \ldots, l$, hence only if $\phi(x)$ indeed contains all the degrees in J_0, J_0' from the 0th to the lth. By (45) it follows

$$q \geq k - \frac{1}{2} + \frac{k - \frac{1}{2} - (1 + 2 + \cdots + q - 1)}{2n - 1},$$

hence by (42) and (46)

$$q \geq k - \frac{1}{2} - \frac{k^2 - 3k + 1}{2(2n - 1)} > k - 1,$$

hence by (46)

$$q = k. \tag{47}$$

Thus all the degrees $0, \ldots, l$ in J_0, J_0' indeed appear in $\phi(x)$; one can identify the functions $\omega_0, \ldots, \omega_q$ with $t_1, \ldots, t_k$ and the polynomials $\tau_{a1}, \ldots, \tau_{aq}$ with the polynomials $\sigma_{a1}, \ldots, \sigma_{ak}$.

Like in the introduction one deduces from the sth derivative of the equation (44) that among the $k + s$ linear forms

$$\sigma_{a1}(\xi)t_1 + \cdots + \sigma_{ak}(\xi)t_k, \qquad (a = 0, \ldots, k + s - 1)$$

of the k variables $t_1, \ldots, t_k$ at least k are independent of each other. From (45) and (47) one gets

$$k + s \leq k + \delta = k(2n - 1) + 1 + 2 + \cdots + k - (2k - 1)n < n + k^2,$$

which completes the proof. □

§3. The transcendence of $J_0(\xi)$.

Let ξ be a non-zero number. Choose k numbers $h_1, \ldots, h_k$ from $0, 1, \ldots, n + k^2 - 1$, which is possible by Theorem 8, so that the k linear forms

$$\phi_\nu = \sigma_{h_\nu 1}(\xi)t_1 + \cdots + \sigma_{h_\nu k}t_k, \qquad (\nu = 1, \ldots, k)$$

are linearly independent.

Let $g(y, z)$ be a polynomial in the variables y, z of degree $p \leq l$ with rational integer coefficients that are not all 0. Let G be an upper bound for the absolute value of the coefficients of $g(y, z)$. Put $l - p = r$ and form the $\frac{(r+1)(r+2)}{2} = v$

polynomials $y^\rho z^{\sigma-\rho} g(y,z)$ with $\rho = 0,\dots,\sigma$ and $\sigma = 0,\dots,r$, whose degrees $p+\sigma$ are all $\le l$. For the special values $y = J_0(x), z = J_0'(x)$ these polynomials are linear and homogeneous in the monomials $J_0^\kappa J_0'^{\lambda-\kappa}$ $(\kappa = 0,\dots,\lambda; \lambda = 0,\dots,l)$, hence in $t_1,\dots,t_k$; one gets in this way v new linear forms in $t_1,\dots,t_k$, say $\psi_1,\dots,\psi_v$, whose coefficients are rational integers and are in absolute value $\le G$. The v polynomials $y^\rho z^{\sigma-\rho} g(y,z)$ are linearly independent which implies that the same holds for $\psi_1,\dots,\psi_v$. Among the k linearly independent forms ϕ_ν $(\nu = 1,\dots,k)$ one chooses $k - v = w$ suitable forms, say ϕ_ν for $\nu = 1,\dots,w$, so that the k forms

$$\psi_1,\dots,\psi_v \text{ and } \phi_1,\dots,\phi_w \tag{48}$$

are linearly independent.

One has now to repeat the arguments of the introduction, which led to (2), in a slightly more general form. Associate t_1 to the monomial $J_0^0 J_0'^0 = 1$. Let Δ be the determinant of the system formed by the k forms (48) and let $\Gamma_1,\dots,\Gamma_v$ and $B_1,\dots,B_w$ the determinants of the submatrices corresponding to the elements of the first column of Δ. Then

$$\Delta = \Gamma_1\psi_1 + \cdots + \Gamma_v\psi_v + B_1\phi_1 + \cdots + B_w\phi_w. \tag{49}$$

From Theorem 7 it was deduced that the rational integer coefficients of the polynomials $\sigma_{ab}(x)$ for $a = 0,\dots,n+k^2-1$ and $b = 1,\cdots,k$ have order $(n!)^{3+\varepsilon}$. The polynomial σ_{ab} has degree $2n+a-1 \le 3n+k^2-2$. The determinant Δ is a polynomial in ξ of degree $w(3n+k^2-2)$, the coefficients of which are rational integers having order $(n!)^{3w+\varepsilon}G^v$ in n and G. Moreover, when ξ is fixed, the determinants $\Gamma_1,\dots,\Gamma_v$ are of order $(n!)^{3w+\varepsilon}G^{v-1}$ and the determinants $B_1,\dots,B_w$ are of order $(n!)^{3(w-1)+\varepsilon}G^v$. Since $\phi_1,\dots,\phi_w$ are bounded from above by (41) times $n!$, the numbers $\phi_1,\dots,\phi_w$ are of order $(n!)^{3+\varepsilon-(2k-1)}$ for ξ fixed.

The right-hand side of (49) therefore is

$$O\left((n!)^{3w+\varepsilon}G^v\left\{\frac{|g(J_0(\xi),J_0'(\xi))|}{G} + (n!)^{1-2k}\right\}\right). \tag{50}$$

Now take for ξ an algebraic number of degree m. One chooses a natural number c so that $c\xi$ is integral. Then $c^{w(3n+k^2-2)}\Delta$ is an integer of the number field associated to ξ; since it is not 0, its norm in absolute value is ≥ 1. Taking into account the estimate for the coefficients of Δ, it follows from (49) and (50) that

$$1 < KG^{vm}(n!)^{3wm+\varepsilon}\left\{\frac{|g(J_0(\xi),J_0'(\xi))|}{G} + (n!)^{1-2k}\right\}, \tag{51}$$

where K does not depend on n and is ≥ 1. Take now $\varepsilon = 1$ and $r = 4pm$, then because of $l = p + r$

$$\begin{aligned} 2k - 2 &= (p+r+1)(p+r+2) - 2 > p^2(4m+1)^2 > 8p^2m(2m+1), \\ v &= \frac{1}{2}(r+1)(r+2) < 2p^2(2m+1)^2, \\ w &= \frac{1}{2}(p+r+1)(p+r+2) - \frac{1}{2}(r+1)(r+2) \\ &= \frac{1}{2}p(p+2r+3) \leq 2p^2(2m+1). \end{aligned}$$

Moreover, one chooses n to be the *smallest* natural number satisfying (42) and the condition

$$n! > 2KG^{2m+1}.$$

Then one has

$$\begin{aligned} (n!)^{2k-1} &> (n!)^{1+8p^2m(2m+1)} > (n!)^{1+6p^2m(2m+1)} 2KG^{2p^2m(2m+1)^2} \\ &\geq 2KG^{vm}(n!)^{3wm+\varepsilon}, \end{aligned}$$

therefore by (51)

$$\left|g(J_0(\xi), J_0'(\xi))\right| > G(n!)^{1-2k}.$$

By observing that $n!$ has order $G^{2m+1} \log G$ and that

$$(2k-1)(2m+1) \leq 3m\{(4pm+p+1)(4pm+p+2) - 1\} \leq 123p^2m^3,$$

one gets

Main Theorem. *Let ξ be a non-zero algebraic number of degree m. Let $g(y, z)$ be a polynomial of degree p in y and z with rational integer coefficients, which are not all 0 and in absolute value $\leq G$. Then for a certain positive number c, only depending on ξ and p, the inequality*

$$\text{(52)} \qquad \left|g(J_0(\xi), J_0'(\xi))\right| > cG^{-123p^2m^3}$$

holds. In particular, the numbers $J_0(\xi)$ and $J_0'(\xi)$ are not related by an algebraic equation with rational coefficients and, especially, the number $J_0(\xi)$ is transcendental.

By the formulation chosen to express the main theorem the negative assertion that the numbers $J_0(\xi)$ and $J_0'(\xi)$ are algebraically independent for algebraic $\xi \neq 0$ got a positive turnaround; namely one now has a positive bound for the distance to 0 of the value of an arbitrary polynomial in $J_0(\xi)$ and $J_0'(\xi)$ with rational coefficients. Thanks to this estimate it is possible to actually *calculate* with the transcendental numbers $J_0(\xi)$, $J_0'(\xi)$ in the same way as one does with algebraic numbers; because one can indeed *decide* how a given algebraic expression in J_0, J_0' with algebraic coefficients is located in relation to a given rational number.

For the constant c in (52) one could easily give an explicit expression in terms of ξ and p, and the exponent $123p^2m^3$ can be reduced, when one makes the estimates sharper. In case $p = 1$ it is possible to give the "exact" exponent. The resulting arithmetic theorem has then again an algebraic counterpart, on which more will be said below.

There are three polynomials of degree n, say $f(x), g(x), h(x)$, so that the power series for

$$f(x)J_0(x) + g(x)J_0'(x) + h(x) = R(x)$$

starts with the power x^{3n+2} or an even bigger one and so that f, g, h do not vanish identically. Then it holds

$$(f' - g)J_0 + \left(g' + f - \frac{g}{x}\right) J_0' + h' = R'$$

$$\left(f'' - f - 2g' + \frac{g}{x}\right) J_0 + \left(2f' - \frac{f}{x} - g + g'' - \frac{2g'}{x} + \frac{2g}{x^2}\right) J_0' + h'' = R'',$$

which will for brevity be denoted by

$$f_1 J_0 + g_1 J_0' + h_1 = R'$$

$$f_2 J_0 + g_2 J_0' + h_2 = R''.$$

Since J_0 is even and J_0' is odd, the polynomials f, h can chosen to be either both even or both odd, and in the latter case g is odd. In both cases fg is divisible by x. Thus the expressions

$$fg_1 - gf_1, \quad x(f_2 g - g_2 f), \quad x^2(f_1 g_2 - g_1 f_2)$$

are polynomials. The determinant

$$\text{(53)} \qquad \begin{vmatrix} f & g & h \\ f_1 & g_1 & h_1 \\ f_2 & g_2 & h_2 \end{vmatrix} = (f_1 g_2 - g_1 f_2)R + (f_2 g - g_2 f)R' + (fg_1 - gf_1)R''$$

is therefore both an entire function, divisible by x^{3n}, and a rational function of degree $3n$. Its value is therefore γx^{3n}, and the constant γ is different from 0 by Theorem 4. From (53) it follows that R indeed starts with the $(3n+2)$th power of x and not with a larger one. Therefore the polynomials f, g, h are uniquely determined up to a common constant factor. Moreover

$$J_0 : J_0' : 1 = (gh_1 - hg_1 + Rg_1 - R'g) : (hf_1 - fh_1 - Rf_1 + R'f) : (fg_1 - gf_1)$$

and this gives an approximation of J_0 and J_0' through rational functions with the same denominator. Probably these generalized continued fraction expansions that appear here may also be of relevance for other linear differential equations.

The statement that the expression $f J_0 + g J_0' + h$ is divisible by x^{3n+2}, but not by x^{3n+3}, where f, g, h are polynomials of degree n, has the following algebraic counterpart:

Let r be a non-zero rational number. Let a, b, c be three rational integer numbers, whose absolute values have the positive maximum M. Then one has

$$\left|a J_0(r) + b J_0'(r) + c\right| > c_1 M^{-2-\varepsilon}, \tag{54}$$

where ε is an arbitrary positive number and $c_1 > 0$ only depends on r and ε.

Here the exponent $-2-\varepsilon$ is best-possible up to the arbitrarily small quantity ε; because for any choice of the real numbers ρ, ς, τ the inequality $|a\rho + b\varsigma + c\tau| \leq (|\rho| + |\varsigma| + |\tau|)M^{-2}$ has always infinitely many solutions in rational integer numbers a, b, c.

The proof of (54) is given exactly in the same way as the one that led to (52), by just refining the estimates a bit. Also more generally it follows:

Let $r \neq 0$ be rational. Let P be a polynomial in $J_0(r)$, $J_0'(r)$ of degree p. Let its $\frac{1}{2}(p+1)(p+2) = q$ coefficients be rational integers and in absolute value $\leq M$. Then

$$\left|P(J_0(r), J_0'(r))\right| > c_2 M^{1-q-\varepsilon} \tag{55}$$

for any $\varepsilon > 0$ and for a $c_2 > 0$ only depending on r, p, ε.

On the other hand one can determine the coefficients of P in infinitely many ways so that

$$\left|P(J_0(r), J_0'(r))\right| < c_3 M^{1-q}$$

holds, for a certain value c_3 only depending on r and p. The inequality (55) thus expresses how much the values $J_0(r)$ and $J_0'(r)$ oppose to be related by an algebraic equation.

As in the previously considered case $p = 1$, there is an algebraic counterpart to (55), which shows how much the functions $J_0(x)$ and $J_0'(x)$ oppose to be related by an algebraic equation with coefficients that are polynomials in x; this is then a refinement of Theorem 2 which was saying that $J_0(x)$ does not satisfy any differential equation of first order.

§4. Further applications of the method.

I.

The E-function

$$K_\lambda(x) = \Gamma(\lambda+1)\left(\frac{x}{2}\right)^{-\lambda} J_\lambda(x) = \sum_{n=0}^{\infty} \frac{(-1)^n}{n!(\lambda+1)\cdots(\lambda+n)}\left(\frac{x}{2}\right)^{2n}$$

with rational $\lambda \neq -1, -2, \ldots$ can be analyzed exactly in the same way as it was done with $K_0 = J_0$. One obtains also here the result that for no algebraic $\xi \neq 0$ an algebraic relation with rational coefficients relating the numbers $K_\lambda(\xi)$ and $K'_\lambda(\xi)$ can exist; an exception is only given by the case that λ is half of an odd number. In particular the non-zero roots of the BESSEL functions $J_\lambda(x)$ for rational λ are always transcendental, and this also holds by LINDEMANN's theorem in the case when 2λ is odd.

From the well-known relations

$$J'_\lambda = \frac{\lambda}{x} J_\lambda - J_{\lambda+1},$$
$$\frac{J_{\lambda-1}}{J_\lambda} = \frac{2\lambda}{x} - \cfrac{1}{\frac{2\lambda+2}{x} - \cdots} \tag{56}$$

it follows that the continued fraction

$$i\frac{J_{\lambda-1}(2ix)}{J_\lambda(2ix)} = \frac{\lambda}{x} + \cfrac{1}{\frac{\lambda+1}{x} + \cfrac{1}{\frac{\lambda+2}{x} + \cdots}}$$

represents a transcendental number for rational λ and algebraic $x \neq 0$. For odd values of 2λ this is contained in LINDEMANN's theorem.

This implies the transcendence of the continued fraction

$$r_1 + \cfrac{1}{r_2 + \cfrac{1}{r_3 + \cdots}}$$

for rational numbers $r_1, r_2, r_3, \ldots$ that belong to an arithmetic progression of first order.

As a special case the continued fraction

$$1 + \cfrac{1}{2 + \cfrac{1}{3 + \cdots}}$$

is transcendental.

II.

It is well known that the functions $J_\lambda(x)$ and $J_{\lambda+1}(x)$ do not have non-zero roots in common, which follows from (56). It will be shown that also J_λ and $J_{\lambda+n}$ for $n = 2, 3, \ldots$ do not have common roots $\neq 0$, provided that λ is rational and that for negative rational integers λ the value $n = -2\lambda$ is avoided. Namely, by (56)

$$J_{\lambda+n} = PJ_\lambda + QJ'_\lambda,$$

where P and Q are rational functions in x with rational coefficients. If Q were identically 0, it would follow that with $\lambda + n = \mu$

$$\frac{d^2(PJ_\lambda)}{dx^2} + \frac{1}{x}\frac{d(PJ_\lambda)}{dx} + \left(1 - \frac{\mu^2}{x^2}\right)PJ_\lambda = 0,$$

thus

(57) $$\left(P'' + \frac{P'}{x} + \frac{\lambda^2 - \mu^2}{x^2}P\right)J_\lambda + 2P'J'_\lambda = 0.$$

If now 2λ is not an odd number, then by Theorem 2 it follows that

$$P' = 0, \quad \lambda^2 = \mu^2, \quad n = -2\lambda, \quad \lambda \text{ integral},$$

and this is the trivial exception $(-1)^\lambda J_\lambda = J_{-\lambda}$. If 2λ is odd, it is well known that

(58) $$J_\lambda = ae^{ix} + be^{-ix}$$

with rational functions a and b, both not identically 0, so

(59) $$J'_\lambda = (a' + ia)e^{ix} + (b' - ib)e^{-ix},$$

and the function $(a' + ia)b - a(b' - ib) = 2iab + ba' - ab'$ is not identically 0 since it has the same degree as ab; from (57), (58) and (59) it follows again that $P' = 0, \lambda^2 = \mu^2$; therefore $J_{-\lambda}$ and J_λ would be proportional, a contradiction.

Hence, Q is not identically equal to 0. Let now $\alpha \neq 0$ be a root of J_λ, implying that this is a simple root, thus $J'_\lambda(\alpha) \neq 0$. Since α is transcendental, $P(\alpha)$ is finite and $Q(\alpha) \neq 0$, thus

$$J_{\lambda+n}(\alpha) = Q(\alpha)J'_\lambda(\alpha) \neq 0.$$

III.

Easier than for $J_0(x)$ is to handle the solutions of

(60) $$y' + \left(\frac{\lambda}{x} - 1\right)y = \frac{\lambda}{x}, \quad \lambda \neq -1, -2, \ldots$$

that is

(61) $$y = x^{-\lambda}e^x\left(\int_0^x \lambda t^{\lambda-1}e^{-t}dt + c\right).$$

For the value 0 of the integration constant c one has

$$y_0 = 1 + \frac{x}{\lambda+1} + \frac{x^2}{(\lambda+1)(\lambda+2)} + \cdots .$$

The transcendence of y_0 follows for any rational λ and any algebraic $x \neq 0$. Consequently, the roots of the function

$$\int_0^1 t^{\lambda-1} e^{-tx}\, dt \tag{62}$$

are transcendental. This includes for $\lambda = 1$ the transcendence of π, and for $x = 1$ the irrationality of the roots of the "incomplete" Gamma function

$$\int_0^1 t^{x-1} e^{-t}\, dt.$$

For algebraic c and rational λ the right-hand side of (61) is also an E-function. Therefore the expression (62) for rational λ and algebraic $x \neq 0$ is a transcendental number, hence also

$$\int_0^{-x} e^{-t^{\lambda}}\, dt.$$

IV.

The solution of (60) is included in the solutions of

$$xy'' + (\lambda - x)y' - \kappa y = 0;$$

one of which is the integral

$$\begin{aligned} y &= 1 + \frac{\kappa}{\lambda}\frac{x}{1!} + \frac{\kappa(\kappa+1)}{\lambda(\lambda+1)}\frac{x^2}{2!} + \cdots \\ &= \int_0^1 t^{\kappa-1}(1-t)^{\lambda-\kappa-1} e^{tx}\, dt : \int_0^1 t^{\kappa-1}(1-t)^{\lambda-\kappa-1}\, dt. \end{aligned}$$

If κ and λ are rational, these functions can be treated in the same way as $J_0(x)$.

V.

By slightly generalizing the previous considerations it is possible to prove the following theorem, that contains both the main theorem on $J_0(x)$ and the general theorem by LINDEMANN:

Let $\xi, \alpha_1, \ldots, \alpha_n$ be algebraic numbers with $\alpha_1, \ldots, \alpha_n$ pairwise distinct and with $\xi \neq 0$. Let $P_1(x, y), \ldots, P_n(x, y)$ be polynomials with algebraic coefficients, not all 0. Then it follows that the number

$$P_1(J_0(\xi), J_0'(\xi))e^{\alpha_1} + \cdots + P_n(J_0(\xi), J_0'(\xi))e^{\alpha_n} \neq 0. \tag{63}$$

In other words: No non-trivial algebraic relations exist between $J_0(\xi)$, $J_0'(\xi)$ and values of the exponential function for algebraic arguments. In particular, the number $J_0(\xi)$ is transcendental over the field of e.

For the proof one needs a generalization of Theorem 2; one has to show that the function

$$P_1(J_0(\xi x), J_0'(\xi x))e^{\alpha_1 x} + \cdots + P_n(J_0(\xi x), J_0'(\xi x))e^{\alpha_n x}$$

does not vanish identically in x. This can be done with the method used in the proof of Theorem 2; compare with VI.

It is also possible to obtain an explicit positive lower bound for the absolute value of the left-hand side of (63). In particular, this is valid for the special case of HERMITE's theorem:

Let $m_0, \ldots, m_n$ be rational integer numbers with the maximum of their absolute values being $M > 0$. Then one has

$$\left| m_0 e^n + m_1 e^{n-1} + \cdots + m_{n-1} e + m_n \right| > c_4 M^{-n-\varepsilon},$$

where $c_4 > 0$ and only depends on ε and n. On the other hand, again

$$\left| m_0 e^n + \cdots + m_n \right| < c_5 M^{-n}$$

for a suitable c_5 has infinitely many solutions in rational integers $m_0, \ldots, m_n$.

There is also an analogous algebraic theorem:

Let $f_0(x), \ldots, f_n(x)$ be polynomials of degree h and not all identically 0, then the function

$$f_0(x)e^{nx} + f_1(x)e^{(n-1)x} + \cdots + f_{n-1}(x)e^x + f_n(x)$$

vanishes at $x = 0$ at most with order $(n+1)(h+1)-1$, and this order of vanishing is indeed attained by suitable polynomials $f_0, \ldots, f_n$, which of course is trivial.

For $n = 1$ both statements follow from the continued fraction expansion for e^x. These assertions express the fact, how much the number e and the function e^x oppose to be algebraic.

For π one obtains the inequality

$$\left| m_0 \pi^n + m_1 \pi^{n-1} + \cdots + m_n \right| > c_6 M^{-M^{\varepsilon}}$$

for arbitrarily small positive ε and a positive c_6 only depending on ε and n.

VI.

The more general question about the algebraic independence of the values of $J_0(\xi)$ for different algebraic values of $\xi^2 \neq 0$ can be answered also, so that one has the same information about $J_0(\xi)$ that one has about the exponential e^x thanks to LINDEMANN's theorem. Actually, it is possible to show:

Let $\xi_1^2, \ldots, \xi_n^2$ be pairwise distinct non-zero algebraic numbers. Then the $2n$ numbers $J_0(\xi_1)$, $J_0'(\xi_1), \ldots, J_0(\xi_n)$, $J_0'(\xi_n)$ are pairwise algebraically independent over the field of rational numbers.

More generally this theorem also holds for K_λ instead of J_0, where λ denotes a rational number that is not half of an odd number. And even this assertion can be further generalized by considering different values of λ simultaneously. In this way one obtains the following statement:

Let $\lambda_1, \ldots, \lambda_m$ be rational numbers; let none of the numbers $2\lambda_1, \ldots$ $\ldots, 2\lambda_m$ be odd, let none of the sums $\lambda_k + \lambda_l$ and of the differences $\lambda_k - \lambda_l$ ($k = 1, \ldots, m$; $l = 1, \ldots, m$; $k \neq l$) be integral. Let $\xi_1^2, \ldots, \xi_n^2$ be pairwise distinct non-zero algebraic numbers. Then no algebraic relation with rational coefficients exists involving the $2mn$ quantities $K_\lambda(\xi), K_\lambda'(\xi)$ ($\lambda = \lambda_1, \lambda_2, \ldots, \lambda_m$; $\xi = \xi_1, \xi_2, \ldots, \xi_n$).

The proof is done exactly in the same way as in the detailed given proof of the special case $m = 1, \lambda_1 = 0, n = 1$. It suffices to deal with the algebraic part of the proof, which leads to the generalization of Theorem 3, namely to

Theorem 9. *Let $\lambda_1, \ldots, \lambda_m$ be arbitrary numbers, none of which being half of an odd number and such that no two of them have rational integer sum or difference. Moreover, let the numbers $\xi_1^2, \ldots, \xi_n^2$ be pairwise distinct and non-zero. Let J_λ and Y_λ be linearly independent solutions of* BESSEL*'s differential equation. Then the $3mn$ functions $J_\lambda(\xi x)$, $J_\lambda'(\xi x)$, $Y_\lambda(\xi x)$ ($\lambda = \lambda_1, \ldots, \lambda_m$; $\xi = \xi_1, \ldots, \xi_n$) are algebraically independent over the field of rational functions of x.*[4]

Proof. The BESSEL function with argument ξx satisfies the differential equation

$$y'' + \frac{1}{x} y' + \left(\xi^2 - \frac{\lambda^2}{x^2}\right) y = 0. \tag{64}$$

Let u and v be two linearly independent solutions. The solutions corresponding to the mn pairs ξ, λ are denoted, in some order, by u_1, v_1; u_2, v_2; ... One has to prove that no algebraic relation with constant coefficients exists involving the $3mn + 1$ functions $x, u_1, u_1', v_1, u_2, u_2', v_2, \ldots$. For $m = 1, n = 1$ this follows by Theorem 3. Let now $mn > 1$ and let the number r of triples u, u', v that really appear in the algebraic relation be smallest possible; furthermore, for this r, assume the number s of really appearing functions v be smallest possible. Hence, one has a relation involving $u_1, u_1', u_2, u_2', \ldots, u_r, u_r'$ and

[4] FOOTNOTE BY THE EDITORS: Compare this result with the more general 'Siegel-Shidlovsky theorem' (see e.g. [A. Baker, Transcendental Number Theory, Cambridge University Press, Cambridge, 1975]). Further generalisations of the stated results appear e.g. in [Y. André, Séries Gevrey de type arithmétique. II. Transcendance sans transcendance, Ann. of Math. (2) **151** (2000), 741–756] and in [D. Bertrand, On André's proof of the Siegel-Shidlovsky theorem, Colloque Franco-Japonais: Théorie des Nombres Transcendants (Tokyo, 1998), 51–63, Sem. Math. Sci., 27, Keio Univ., Yokohama, 1999].

$v_1, \ldots, v_s$ with $s \leq r$, whose coefficients are polynomials in x; and in each such relation one of the functions u_r, u'_r, v_r and, in case $s > 0$, the functions v_s really appear.

First one shows now that s has the value 0. For this one just has to repeat the arguments from the proof of Theorem 3; equality (18) has thereby to be replaced by an equality involving $x, u_2, u'_2, \ldots, u_r, u'_r$ and $v_1, \ldots, v_s$ and $\lambda u_1 + \mu v_1, \lambda u'_1 + \mu v'_1$, where λ and μ denote suitable constants. Replacing $\lambda u_1 + \mu v_1$ by u_1, one is led to an equation with the same r and a smaller s. Therefore $s = 0$.

Next it shall be proved that also a relation exists, in which u'_1 does not appear. The deduction is done as in the proof of Theorem 2. If $\Phi = 0$ is an irreducible algebraic equality in $x, u_1, u'_1, \ldots, u_r, u'_r$ which contains u'_1, then from $\Phi = 0$ and from $\frac{d\Phi}{dx} = 0$, using (64), it follows that one can either eliminate u'_1 or obtain, in analogy to (11), an equality

$$(65) \qquad \begin{aligned} &Q(\lambda_1 u_1 + \mu_1 v_1, \lambda_1 u'_1 + \mu_1 v'_1, \lambda_2 u_2 + \mu_2 v_2, \ldots) \\ &\quad = k(\lambda_1, \mu_1, \lambda_2, \mu_2, \ldots) x^b e^{ax - \frac{c}{x}} \end{aligned}$$

where k is a polynomial with constant coefficients, that is homogeneous in any of the r pairs λ_a, μ_a $(a = 1, \ldots, r)$, and $Q(u_1, u'_1, \ldots)$ denotes a polynomial in $x, u_1, u'_1, \ldots$ that is homogeneous in any of the pairs u_a, u'_a $(a = 1, \ldots, r)$. One defines $\lambda_a : \mu_a = \rho_a$ and assumes for these parameters the condition $k = 0$, then $\rho_1 = \rho$ is an algebraic function in $\rho_2, \ldots, \rho_r$. If ρ were identically constant, then the left-hand side of (65) would, for this value of $\rho = \lambda_1 : \mu_1$, identically vanish in $\lambda_2, \mu_2, \ldots, \lambda_r, \mu_r$; but in this case the coefficient of any monomial in $u_2, u'_2, \ldots, u_r, u'_r$ in $Q(\lambda_1 u_1 + \mu_1 v_1, \lambda_1 u'_1 + \mu_1 v'_1, u_2, u'_2, \ldots, u_r, u'_r)$ would be 0 and this would give an equality in $\lambda_1 u_1 + \mu_1 v_1, \lambda_1 u'_1 + \mu_1 v'_1, x$ contradicting Theorem 3. Hence, ρ is not constant for varying $\rho_2, \ldots, \rho_r$. The function $(\rho u'_1 + v'_1)/(\rho u_1 + v_1)$ is an algebraic function in x and in the $r - 1$ functions $(\rho_2 u'_2 + v'_2)/(\rho_2 u_2 + v_2), \ldots, (\rho_r u'_r + v'_r)/(\rho_r u_r + v_r)$ and shall be denoted by A. One chooses for $\rho_2, \ldots, \rho_r$ four arbitrary sets of values, which shall be indicated by the upper indices I, ..., IV; the corresponding values of ρ and A shall be denoted by attaching the appropriate indices. From

$$(66) \qquad \frac{\rho u'_1 + v'_1}{\rho u_1 + v_1} = A$$

it follows that

$$(67) \qquad \frac{\rho^{\mathrm{I}} - \rho^{\mathrm{III}}}{\rho^{\mathrm{I}} - \rho^{\mathrm{IV}}} : \frac{\rho^{\mathrm{II}} - \rho^{\mathrm{III}}}{\rho^{\mathrm{II}} - \rho^{\mathrm{IV}}} = \frac{A^{\mathrm{I}} - A^{\mathrm{III}}}{A^{\mathrm{I}} - A^{\mathrm{IV}}} : \frac{A^{\mathrm{II}} - A^{\mathrm{III}}}{A^{\mathrm{II}} - A^{\mathrm{IV}}},$$

and this identically holds in the $4(r-1)$ variables $\rho_2^{\mathrm{I}}, \ldots, \rho_r^{\mathrm{I}}, \ldots, \rho_2^{\mathrm{IV}}, \ldots$ $\ldots, \rho_r^{\mathrm{IV}}$ if ρ is assumed to be the algebraic function in $\rho_2, \ldots, \rho_r$ obtained by setting $k = 0$. By (14) for a constant $a \neq 0$

$$(68) \qquad u'v - v'u = \frac{a}{x};$$

substituting in (67) the values for $v'_2, \ldots, v'_r$ obtained by the last relation, one gets an algebraic relation between x and just $r-1$ triples $u_2, u'_2, v_2, \ldots, u_r, u'_r, v_r$, *i.e.* an equation holding identically in $u_2, u'_2, v_2, \ldots, u_r, u'_r, v_r$. But now A only depends on the ratios $\frac{\rho_2 u'_2 + v'_2}{\rho_2 u_2 + v_2}, \ldots$, hence, only on the ratios $u'_2 : v'_2 : u_2 : v_2, \ldots$; (68) already means that for arbitrary $u' : v' : u : v$ the proportionality factor of u', v', u, v has been fixed, consequently (67) holds identically in the $4(r-1)$ variables $u_2, u'_2, v_2, v'_2, \ldots$.

Therefore one can replace the function $\frac{\rho_2 u'_2 + v'_2}{\rho_2 u_2 + v_2}$, which is linear in ρ_2, by

$$\frac{\rho_2 - \rho_2^{\mathrm{III}}}{\rho_2 - \rho_2^{\mathrm{IV}}} : \frac{\rho_2^{\mathrm{II}} - \rho_2^{\mathrm{III}}}{\rho_2^{\mathrm{II}} - \rho_2^{\mathrm{IV}}},$$

and analogously for the other indices $3, \ldots, r$. Then, the arguments of A^{II} all become equal to 1, those of A^{III} all equal to 0, those of A^{IV} all equal to ∞, and those of A^{I} become the cross-ratios of $\rho_a^{\mathrm{I}}, \rho_a^{\mathrm{II}}, \rho_a^{\mathrm{III}}, \rho_a^{\mathrm{IV}}$ for $a = 2, \ldots, r$ respectively. By (67) the cross-ratio of $\rho^{\mathrm{I}}, \rho^{\mathrm{II}}, \rho^{\mathrm{III}}, \rho^{\mathrm{IV}}$ is then an algebraic function of these $r-1$ cross-ratios, and on the other hand ρ^{I} is an algebraic function in $\rho_2^{\mathrm{I}}, \ldots, \rho_r^{\mathrm{I}}$ only, $\ldots, \rho^{\mathrm{IV}}$ being the same algebraic function in $\rho_2^{\mathrm{IV}}, \ldots, \rho_r^{\mathrm{IV}}$ only.

If for a given function $f(x)$, differentiable and not identically constant in an interval, the cross-ratio $\Delta = \frac{f(x_1)-f(x_3)}{f(x_1)-f(x_4)} : \frac{f(x_2)-f(x_3)}{f(x_2)-f(x_4)}$ is a function in $\frac{x_1-x_3}{x_1-x_4} : \frac{x_2-x_3}{x_2-x_4} = D$, then, letting x_1 and x_2 converge to x_3 for a fixed value of D, the relation

$$\Delta : D = \lim \frac{f(x_1) - f(x_3)}{x_1 - x_3} : \frac{f(x_2) - f(x_3)}{x_2 - x_3} = 1$$

follows; hence $\Delta = D$.

In (67) the right-hand side is therefore the first argument of A^{I}, moreover $A^{\mathrm{II}}, A^{\mathrm{III}}, A^{\mathrm{IV}}$ are algebraic functions in x. Consequently, A is a linear function of $\frac{\rho_2 u'_2 + v'_2}{\rho_2 u_2 + v_2}$ whose coefficients are algebraic functions in x. By (66), ρ is a linear function in ρ_2. Replacing u_1 and v_1 with suitable homogeneous linear combinations of u_1 and v_1 with constants coefficients, it follows $\rho_2 = \rho$. From

$$\frac{\rho u'_1 + v'_1}{\rho u_1 + v_1} = \frac{\alpha \frac{\rho u'_2 + v'_2}{\rho u_2 + v_2} + \beta}{\gamma \frac{\rho u'_2 + v'_2}{\rho u_2 + v_2} + \delta}$$

with algebraic $\alpha, \beta, \gamma, \delta$ it follows by differentiating w.r.t. ρ and setting $\rho = \infty$ that

$$\frac{u'_1 v_1 - v'_1 u_1}{u_1^2} = (\alpha\delta - \beta\gamma) \frac{u'_2 v_2 - v'_2 u_2}{(\gamma u'_2 + \delta u_2)^2},$$

hence by (68)

$$u_1 = p u_2 + q u'_2$$

with p and q algebraic functions in x only. This equality does not involve u'_1.

It follows that there is an equation involving $x, u_2, u_2', \ldots, u_r, u_r'$ and u_1. Denote it by $\psi = 0$ and assume it to be irreducible. From $\frac{d\psi}{dx} = 0$ one sees that u_1' is a rational function in $x, u_2, u_2', \ldots, u_r, u_r'$ and u_1, and by differentiation u_1'' is also such a function. Substituting u_1 for y in the left-hand side of (64), the resulting expression must vanish; it is again an equation $\chi = 0$ involving $x, u_2, u_2', \ldots, u_r, u_r'$ and u_1. The elimination of u_1 from $\chi = 0$ and $\psi = 0$ is impossible. Consequently, the equation $\chi = 0$ is satisfied if $u_2, \ldots, u_r$ are replaced by any solutions of the differential equations satisfied by these functions, and u_1 is determined from $\psi = 0$. Writing $\lambda_2 u_2 + \mu_2 v_2, \lambda_2 u_2' + \mu_2 v_2', \lambda_3 u_3 + \mu_3 v_3, \lambda_3 u_3' + \mu_3 v_3' \ldots$ in place of $u_2, u_2', u_3, u_3', \ldots$, then $\lambda_1 u_1 + \mu_1 v_1$ will satisfy the equation $\psi = 0$ for u_1; here λ_1 and μ_1 are viewed as functions in $\lambda_2, \mu_2, \lambda_3, \mu_3, \ldots$. Let

$$u_1 = A(u_2, u_2', \ldots),$$

where A is an algebraic function in its arguments; then

$$\lambda_1 u_1 + \mu_1 v_1 = A(\lambda_2 u_2 + \mu_2 v_2, \lambda_2 u_2' + \mu_2 v_2', \ldots). \tag{69}$$

In particular, replace μ_2 by 1 and choose for λ_2 three arbitrary values $\lambda_2^{\mathrm{I}}, \lambda_2^{\mathrm{II}}, \lambda_2^{\mathrm{III}}$; then A is transformed into $A^{\mathrm{I}}, A^{\mathrm{II}}, A^{\mathrm{III}}$ and the pair λ_1, μ_1 into $\lambda_1^{\mathrm{I}}, \mu_1^{\mathrm{I}}, \ldots$ respectively. Eliminate u_1, v_1 from these three equations; this leads to

$$\rho_1 A^{\mathrm{I}} + \rho_2 A^{\mathrm{II}} + \rho_3 A^{\mathrm{III}} = 0, \tag{70}$$

where ρ_1, ρ_2, ρ_3 are functions in $\lambda_2^{\mathrm{I}}, \lambda_2^{\mathrm{II}}, \lambda_2^{\mathrm{III}}$ independent of x. Eliminating $v_2', v_3', \ldots$ with the help of (68), then (70) transforms into an equation involving $u_2, u_2', v_2, \ldots, u_r, u_r', v_r, x$ and this equation must be identically satisfied in the $3(r-1)+1$ previously mentioned variables. The first two arguments of A are, ignoring for brevity the lower index 2, $\lambda u + v = \xi$ and $\frac{u'}{u}(\lambda u + v) - \frac{a}{xu} = \eta$. One now expands $A(\xi, \eta)$ in powers of $\xi - v = \lambda u$ and obtains

$$A(\xi, \eta) = A(v, \eta) + A_v(v, \eta)\lambda u + \frac{1}{2} A_{vv}(v, \eta)(\lambda u)^2 + \cdots. \tag{71}$$

Moreover, one expands $A\left(v, -\frac{a}{xu}\right)$ in powers of u; let the resulting expansion be

$$A\left(v, -\frac{a}{xu}\right) = c_0 u^{r_0} + c_1 u^{r_1} + \cdots, \qquad (r_0 < r_1 < \ldots) \tag{72}$$

where $c_0, c_1, \ldots$ denote functions in v, none being identically 0. Since (70) holds identically in $u, u', v, \ldots$, one can specialize $u' = 0$; thereby η transforms into $-\frac{a}{xu}$. Substitute the expansion

$$\begin{aligned} A\left(\lambda u + v, -\frac{a}{xu}\right) &= (c_0 u^{r_0} + c_1 u^{r_1} + \cdots) \\ &\quad + \left(\frac{\partial c_0}{\partial v} u^{r_0} + \frac{\partial c_1}{\partial v} u^{r_1} + \cdots\right)\lambda u + \cdots, \end{aligned}$$

obtained from (71) and (72), in (70) and send u to 0. It follows

$$c_0(\rho_1+\rho_2+\rho_3)=0,$$

hence $\rho_1+\rho_2+\rho_3=0$ and $c_1(\rho_1+\rho_2+\rho_3)=0,\ldots$. Let $\frac{\partial c_k}{\partial v}$ be the first not identically vanishing function among $\frac{\partial c_0}{\partial v}, \frac{\partial c_1}{\partial v}, \ldots$, then also

$$\frac{\partial c_k}{\partial v}(\rho_1\lambda^{\mathrm{I}}+\rho_2\lambda^{\mathrm{II}}+\rho_3\lambda^{\mathrm{III}})=0,$$

hence $(\rho_1\lambda^{\mathrm{I}}+\rho_2\lambda^{\mathrm{II}}+\rho_3\lambda^{\mathrm{III}})=0$ and $\frac{\partial c_{k+1}}{\partial v}(\rho_1\lambda^{\mathrm{I}}+\rho_2\lambda^{\mathrm{II}}+\rho_3\lambda^{\mathrm{III}})=0,\ldots$. Let $\frac{\partial^2 c_l}{\partial v^2}$ be the first not identically vanishing function among $\frac{\partial^2 c_k}{\partial v^2}, \frac{\partial^2 c_{k+1}}{\partial v^2}, \ldots$, then also $\rho_1\left(\lambda^{\mathrm{I}}\right)^2+\rho_2(\lambda^{\mathrm{II}})^2+\rho_3(\lambda^{\mathrm{III}})^2=0$, a contradiction if $\lambda^{\mathrm{I}}, \lambda^{\mathrm{II}}, \lambda^{\mathrm{III}}$ are chosen pairwise distinct. Therefore $c_0, c_1, c_2, \ldots$ are either constant or linear in v. Taking (72) into account, it follows

$$A(\xi,\eta)=B(\eta)+\xi C(\eta),$$

where B and C are algebraic functions in η. Now let u' be arbitrary and expand $B(\eta)$ and $C(\eta)$ in powers of $\frac{u'}{u}(\lambda u+v)=\eta+\frac{a}{xu}$. Thereby (70) transforms into

$$\begin{aligned}
&\rho_1\Bigg\{B\left(-\frac{a}{xu}\right)+B'\left(-\frac{a}{xu}\right)\frac{u'}{u}(\lambda^{\mathrm{I}}u+v)\\
&\quad+\frac{1}{2}B''\left(-\frac{a}{xu}\right)\left(\frac{u'}{u}\right)^2(\lambda^{\mathrm{I}}u+v)^2\\
&\quad+\cdots+(\lambda^{\mathrm{I}}u+v)\left(C\left(-\frac{a}{xu}\right)+C'\left(-\frac{a}{xu}\right)\frac{u'}{u}(\lambda^{\mathrm{I}}u+v)+\cdots\right)\Bigg\}\\
&\quad+\rho_2\{\cdots\}+\rho_3\{\cdots\}=0.
\end{aligned}$$

This holds identically in u', so the coefficient of $u':u$ is

$$\rho_1\left\{B'\left(-\frac{a}{xu}\right)(\lambda^{\mathrm{I}}u+v)+C'\left(-\frac{a}{xu}\right)(\lambda^{\mathrm{I}}u+v)^2\right\}+\rho_2\{\cdots\}+\rho_3\{\cdots\}=0,$$

identically in v; consequently, $C'\left(-\frac{a}{xu}\right)=0$, $C(\eta)$ is constant. Furthermore, the coefficient of $(u':u)^2$ is

$$\begin{aligned}
\rho_1B''\left(-\frac{a}{xu}\right)(\lambda^{\mathrm{I}}u+v)^2&+\rho_2B''\left(-\frac{a}{xu}\right)(\lambda^{\mathrm{II}}u+v)^2\\
&+\rho_3B''\left(-\frac{a}{xu}\right)(\lambda^{\mathrm{III}}u+v)^2=0
\end{aligned}$$

identically in v; hence $B''\left(-\frac{a}{xu}\right)=0$, $B(\eta)$ is linear.

It has been shown that A is a linear function of its first two arguments; by (69) also λ_1 and μ_1 are linear functions in λ_2 and μ_2. Set $\lambda_2=0$, $\mu_2=0$, then (69) is not satisfied by the pair of variables u_2, u_2' anymore, hence λ_1 and μ_1 must vanish and every term in $A(u_2, u_2', \ldots)$ has to contain u_2, u_2'. Hence, A is linear and homogeneous in u_2, u_2', and λ_1, μ_1 are linear and homogeneous in λ_2, μ_2. If A would contain a further pair of variables u_3, u_3', then λ_1, μ_1 would also be linear and homogeneous in λ_3, μ_3, say $\lambda_1 = a\lambda_3 + b\mu_3$, $\mu_1 = c\lambda_3 + d\mu_3$. Since $ad - bc$ is a homogeneous quadratic function in λ_2, μ_2, one could find values for $\lambda_2, \mu_2, \lambda_3, \mu_3$ so that λ_1 and μ_1 are 0, but are not all 0 themselves. But in this case (69) would not contain the variables u_1, u_1' anymore.

This leads to an equation

$$u_1 = \alpha u_2 + \beta u_2',$$

where α and β are algebraic functions in x. Substituting this value of u_1 in the differential equation and using the differential equation for u_2 in order to eliminate u_2'' and u_2''', one obtains

$$Ru_2' + Su_2 = 0$$

with

$$R = \beta'' - \frac{1}{x}\beta' + \left(\xi_1^2 - \xi_2^2 + \frac{\lambda_2^2 - \lambda_1^2 + 1}{x^2}\right)\beta + 2\alpha'$$

$$S = \alpha'' + \frac{1}{x}\alpha' + \left(\xi_1^2 - \xi_2^2 + \frac{\lambda_2^2 - \lambda_1^2}{x^2}\right)\alpha - 2\left(\xi_2^2 - \frac{\lambda_2^2}{x^2}\right)\beta' - 2\frac{\lambda_2^2}{x^3}\beta,$$

consequently, $R = 0$, $S = 0$. Here, in contrast to the statement of Theorem 9, ξ_1, λ_1 and ξ_2, λ_2 denote the values of ξ, λ with respect to the functions u_1, u_2 in the differential equation (64).

If one expands α and β in decreasing powers of x

$$\alpha = ax^r + \cdots, \quad \beta = bx^s + \cdots, \quad ab \neq 0;$$

then R and S transform into

$$b(\xi_1^2 + \xi_2^2)x^s + 2arx^{r-1} + \cdots$$

and

$$a(\xi_1^2 + \xi_2^2)x^r - 2b\xi_2^2 sx^{s-1} + \cdots;$$

so $\xi_1^2 = \xi_2^2$. Without loss of generality, let $\xi_1^2 = 1$. From $R = 0$ and $S = 0$, through elimination of α, a fourth order homogeneous linear differential equation for β is obtained:

$$\begin{aligned} x^4\beta'''' &+ 2x^3\beta''' + x^2(4x^2 - 2\lambda_1^2 - 2\lambda_2^2 + 1)\beta'' \\ &+ x(8x^2 + 2\lambda_1^2 + 2\lambda_2^2 - 1)\beta' \\ &+ (\lambda_1 + \lambda_2 + 1)(\lambda_1 + \lambda_2 - 1)(\lambda_1 - \lambda_2 + 1)(\lambda_1 - \lambda_2 - 1)\beta = 0. \end{aligned}$$

The algebraic function β is therefore singular at most at 0 and ∞. Put

$$\beta = x^r(a_0 + a_1x + \cdots + a_lx^l),$$

where $a_0a_l \neq 0$, then comparison of coefficients gives

$$4(r+l)(r+l+1) = 0,$$

hence r must be an integer, and moreover

$$(r-\lambda_1-\lambda_2-1)(r+\lambda_1+\lambda_2-1)(r-\lambda_1+\lambda_2-1)(r+\lambda_1-\lambda_2-1) = 0.$$

Consequently, either $\lambda_1+\lambda_2$ or $\lambda_1-\lambda_2$ is a rational integer, which contradicts the hypothesis. So Theorem 9 is proved. □

The assumptions in Theorem 9 are necessary as is easily seen; e.g. for a rational integer $\lambda_1-\lambda_2$ there exist an algebraic relation between $J_{\lambda_1}, J'_{\lambda_1}, J_{\lambda_2}$ and x, as follows from (56). If one wants to allow for 2λ also odd values, then Theorem 9 needs an extension to which the following remark can be made. In this case $J_\lambda, J'_\lambda, Y_\lambda$ can be expressed algebraically through x and e^{ix} and one has to study whether an expression

$$\Phi = \phi_0 + \phi_1e^{\alpha_1x} + \phi_2e^{\alpha_2x} + \cdots + \phi_ne^{\alpha_nx}$$

can vanish identically in x for pairwise distinct numbers $\alpha_1, \ldots, \alpha_n$ and $\phi_1, \ldots, \phi_n$ polynomials in the BESSEL functions of Theorem 9 and in x. Let n be as small as possible. If $\Phi = 0$, it follows that

$$\frac{d\Phi}{dx} = \frac{d\phi_0}{dx} + \frac{d\phi_1}{dx}e^{\alpha_1x} + \cdots + \frac{d\phi_n}{dx}e^{\alpha_nx} + \alpha_1\phi_1e^{\alpha_1x} + \cdots + \alpha_n\phi_ne^{\alpha_nx} = 0,$$

hence the linear functions Φ and $\frac{d\Phi}{dx}$ in $e^{\alpha_1x}, \ldots, e^{\alpha_nx}$ have to be proportional,

$$\frac{d\log\phi_0}{dx} = \frac{d\log\phi_1}{dx} + \alpha_1 = \cdots = \frac{d\log\phi_n}{dx} + \alpha_n$$

$$\phi_0 = c\phi_1e^{\alpha_1x} \tag{73}$$

hence $n = 1$ has to hold. Replacing x with $2x$ and eliminating e^{α_1x} from the resulting equation and (73), then one obtains a contradiction to Theorem 9. Therefore Φ does not vanish identically in x.

VII.

For power series with *finite* radius of convergence it is not possible to deduce transcendence statements by the method developed above; however, in many cases statements on the irrationality can be achieved.

Instead of the E-functions $\sum_{n=0}^{\infty} \frac{a_n}{b_n} \frac{x^n}{n!}$ consider now power series

$$y = \sum_{n=0}^{\infty} \frac{a_n}{b_n} x^n,$$

in which the factor $n!$ in the denominator is missing, but which satisfy otherwise the same conditions 1., 2., 3. in the definition of the E-functions; the condition about the growth rate of the conjugates of a_n and the least common multiple of $b_1, \ldots, b_n$ shall be replaced with the stronger condition that the growth rate is not larger than a power c^n with suitable constant basis c. Such functions will be called G-functions; the geometric series obviously belongs to this class.[5] Similarly to E-functions, the G-functions form a ring. It is trivial that the derivative of a G-function is again a G-function. From Theorem 5 it follows that the integral $\int_0^x G(t)dt$ is also a G-function.

According to a theorem by EISENSTEIN, every algebraic function regular at $x = 0$ which satisfies an algebraic equation with algebraic coefficients is a G-function. The integral from 0 to x of such an algebraic function is therefore a G-function too. Another example of a G-function is given by the hypergeometric series

$$1 + \frac{\alpha\beta}{\gamma \cdot 1} x + \frac{\alpha(\alpha+1)\beta(\beta+1)}{\gamma(\gamma+1) \cdot 1 \cdot 2} x^2 + \cdots,$$

where α, β, γ are rational.

One can apply to these functions the same considerations that were applied to J_0 before.[6] When differentiating an approximation form in the way done in §2, one has to observe in the present case that the approximation form obtained through iterated differentiation of an approximation form still have a greatest common divisor in their numerical coefficients; only after division by this factor a successful estimate of the rest term becomes possible.

[5] FOOTNOTE BY THE EDITORS: For more on G-functions we refer e.g. to [E. Bombieri, On G-functions, *Recent progress in analytic number theory, Vol. 2 (Durham, 1979)*, 1-67, Academic Press, London-New York, 1981], where (part of) the programme that Siegel asked for in this paper is carried out; see also [Y. André, G-functions and geometry, Aspects of Mathematics, E13, Friedr. Vieweg & Sohn, Braunschweig, 1989].

[6] FOOTNOTE BY THE EDITORS: For recent results on this see [P.-A. Desrousseaux, Periods of abelian varieties, hypergeometric functions and transcendence, Advances in mathematics research. Vol. 9, 157–181, Adv. Math. Res., 9, Nova Sci. Publ., New York, 2009] or [P. Tretkoff, Transcendence of values of transcendental functions at algebraic points (inaugural Monroe H. Martin Lectures and Seminar), Noncommutative geometry, arithmetic, and related topics, 279–295, Johns Hopkins Univ. Press, Baltimore, MD, 2011].

In the investigation of the ABEL integrals $\int_0^x y dt$ it becomes necessary to introduce more general approximation forms. In fact one has to allow as coefficients of the approximation form not only polynomials in the independent variable x alone, but also polynomials in x and y or integral functions belonging to the algebraic function field generated by x and y. This generalisation does not involve any essential difficulties.

The realization of the calculation gives the following result: Let y be an algebraic function in x satisfying an equation with algebraic coefficients. Let $x = 0$ be a regular point of this function. Assume that the ABEL integral $\int_0^x y dt$ is not an algebraic function. Let $\xi \neq 0$ satisfy an algebraic equation of degree l, the coefficients of which are rational integers and in absolute value $\leq M$. Let ϵ be an arbitrary positive number and

$$|\xi| < c_7 e^{-(\log n)^{\frac{1}{2}+\epsilon}}, \tag{74}$$

where c_7 is some positive function of l and ϵ. Then the number $\int_0^\xi y dt$ does not satisfy any algebraic equation of degree l with rational coefficients.

The condition (74) is for instance satisfied for $\xi = \frac{1}{n}$ with n a sufficiently large integer, or for $\xi = (\sqrt{2} - 1)^n$ and n sufficiently large, more generally for every sufficiently large power of any algebraic number inside the unit circle.

Therefore for a sufficiently large integer n the number $\int_0^{\frac{1}{n}} \frac{dx}{\sqrt{1-x^4}}$ is not an algebraic irrationality of degree less than 1000; and more generally, (74) gives infinitely many algebraic ξ of given degree, so that the elliptic integral

$$\int_0^\xi \frac{dx}{\sqrt{4x^3 - g_2 x - g_3}}$$

does not satisfy any algebraic relation of given degree for algebraic values of g_2 and g_3.

Let $x = \eta$ be an algebraic number and a regular point of the algebraic function y. If $\int_0^\eta y dt = \gamma$ satisfies an algebraic equation of degree $\leq l$, then put

$$\int_0^x y(t) dt = \gamma + \int_0^{x-\eta} y(t+\eta) dt$$

and then choose $x = \zeta$ in a neighbourhood of $x = \eta$ so that $\zeta - \eta = \xi$ belongs to the number field given by η and satisfies the inequality (74), and so that the number $\int_0^\xi y(t+\eta) dt$ does not satisfy any equation of degree $\leq l^2$. But then the number $\int_0^\zeta y dt$ will not satisfy any equation of degree l. Thereby it is proved that for every imaginary number field those numbers η from the field, for which the value $\int_0^\eta y dt$ is not an algebraic irrationality of degree $\leq l$, are everywhere dense in the sliced x-plane.

The ABEL integrals, which are algebraic functions in x themselves, obviously do not have the property just mentioned. Therefore an arithmetic property has been found that distinguishes the transcendental ABEL integrals from the algebraic ones.

Applying these remarks to the particular function $\int_0^x \frac{dt}{1+t} = \log(1+x)$, one does not get LINDEMANN's theorem because of the restriction (74). In this way it is not possible to show that the numbers $\log 2, \log 3, \ldots$ are transcendental, it just follows that among them arbitrarily high irrationalities occur, more precisely, it follows that the proportion of the numbers $\log 2, \ldots, \log n$ that are algebraic irrationalities of bounded degree are, for any $\epsilon > 0$, certainly $o(n^\epsilon)$.

Neither can these remarks be used in relation to arithmetical studies of π nor to the periods of ABEL integrals; it is not possible to verify condition (74) using branching theory.

One can formulate the theorem on ABEL integrals also as an assertion concerning their inverse functions. For instance, it follows that the WEIERSTRASS $\wp$-function obtained with algebraic g_2, g_3 does not assume rational integer values for any algebraic irrational argument of degree l which is sufficiently close to the origin.[7]

Analogous considerations can be made for the hypergeometric series for rational α, β, γ. One has to transfer Theorems 1, 2, 3 and, in particular, to identify all hypergeometric functions which satisfy first order algebraic differential equations. This can be done using the method of the proof of Theorem 2. It turns out that only the exceptional cases according to SCHWARZ come into consideration.

For the non-algebraic hypergeometric functions $F(\alpha, \beta, \gamma, x)$ with rational α, β, γ a corresponding irrationality theorem can be shown, similar to the result on ABEL integrals given above; and by the arithmetical property therein the non-algebraic hypergeometric series can be distinguished from the algebraic ones.

An example is provided by the special function

$$F\left(\frac{1}{2}, \frac{1}{2}, 1, x\right) = \sum_{n=0}^{\infty} \binom{2n}{n}^2 \left(\frac{x}{16}\right)^n = \frac{2}{\pi} \int_0^1 \frac{dt}{\sqrt{(1-t^2)(1-xt^2)}}$$
$$= \int_0^1 \frac{dt}{\sqrt{(1-t^2)(1-xt^2)}} : \int_0^1 \frac{dt}{\sqrt{1-t^2}}.$$

One obtains the result that for rational $x = \frac{p}{q}$ its value is irrational provided that

$$0 < \left|\frac{p}{q}\right| < c_8 10^{-\sqrt{\log|q|}},$$

where c_8 denotes a positive constant; and similar theorems hold for higher irrationality. For these values of the modulus the real period of the elliptic integ-

[7] FOOTNOTE BY THE EDITORS: Siegel proved in 1932 the first transcendence results on periods of elliptic functions (cf. [C.L. Siegel, Über die Perioden elliptischer Funktionen, J. Reine Angew. Math. **167** (1932), 62–69]). For surveys of later developments see [T. Schneider, Einführung in die transzendenten Zahlen, Springer, Berlin, 1957] and [D. Masser, Elliptic functions and transcendence, Lecture Notes in Mathematics, Vol. 437. Springer-Verlag, Berlin-New York, 1975].

ral $\int \frac{dt}{\sqrt{(1-t^2)(1-xt^2)}}$ is then incommensurable with π. For the other period no results can be obtained because the corresponding hypergeometric function is logarithmically branched at the origin.

It can be mentioned that numbers like $2^{\sqrt{2}}$ escape from the previous considerations, because the denominators in the binomial series $(1+x)^\lambda = 1+\binom{\lambda}{1}x+\cdots$ grow too fast if λ is an algebraic irrational number.[8]

VIII.

The considerations in VI. can be transferred to G-functions. One then obtains theorems of the sort: If algebraic numbers $\xi_1, \xi_2, \ldots$ satisfy the condition (74), then no algebraic relation of not too large degree involving the values $G_1(\xi_1), G_2(\xi_2), \ldots$ can hold. Thereby one obtains results concerning the logarithmic function that are not contained in LINDEMANN's theorem. In this way it follows that the numbers

$$\frac{\log 2}{\log 3}, \frac{\log 3}{\log 4}, \frac{\log 4}{\log 5}, \ldots$$

contain irrationalities of arbitrarily large degree. So far just the trivial result was known that all these numbers are irrational. Among BRIGG's logarithms of the natural numbers there therefore exist arbitrarily large irrationalities, and this is also true for every other basis of the logarithmic system. More generally, there exist infinitely many positive rational numbers $r_1, \ldots, r_n$, so that no algebraic relation among $\log r_1, \ldots, \log r_n$ with rational integer coefficients and bounded degree holds; in particular this holds for n logarithms linearly independent over a given field.

Finally one obtains an approach to arithmetical theorems via SCHWARZ's automorphic functions, by studying the equations that can hold among the values of different hypergeometric functions at algebraic arguments.

[8] FOOTNOTE BY THE EDITORS: This was later proved independently by A.O. Gelfond and Siegel's student T. Schneider as part of their solution of Hilbert's seventh problem.

Part II: On Diophantine equations.

In memoriam ARTHUR SCHOENFLIES.

The mathematical science is much indebted to ANDRÉ WEIL for the proof of an important arithmetic property of algebraic curves. Before formulating WEIL's theorem the following can be mentioned:

Let $f(x, y) = 0$ be the equation of an algebraic curve of genus $p > 0$. Let the coefficients of the polynomial f belong to an algebraic number field $\mathfrak{K}$. A set of p points on the curve with the coordinates x_l, y_l $(l = 1, \ldots, p)$ will be called *rational*, if all symmetric rational combinations in the p pairs x_l, y_l with coefficients in $\mathfrak{K}$ still belong to $\mathfrak{K}$. To every system $\mathfrak{P}$ of p points on the curve a set of p complex numbers $s_1, \ldots, s_p$ can be associated by first considering the p ABEL integrals of the first kind w_k $(k = 1, \ldots, p)$ associated to the curve from the points $x_l^{(0)}, y_l^{(0)}$ of a fixed system $\mathfrak{P}_0$ to the points x_l, y_l of $\mathfrak{P}$ and then adding, for every $k \in \{1, \ldots, p\}$, the values for $l = 1, \ldots, p$; the resulting p integral sums shall be denoted by $s_1, \ldots, s_p$. Assume further that the set $\mathfrak{P}_0$ of points is rational. If $\mathfrak{P}$ is rational too, one calls, for brevity, the system of the p numbers $s_1, \ldots, s_p$ rational. By ABEL's theorem, if $s_1, \ldots, s_p$ and $t_1, \ldots, t_p$ are rational, then $s_1 + t_1, \ldots, s_p + t_p$ and $s_1 - t_1, \ldots, s_p - t_p$ are also rational. Therefore the rational systems $s_1, \ldots, s_p$ form a module.

The theorem of WEIL *says that this module has a finite basis*. The special case $p = 1$ of this theorem has been discovered for the field of rational numbers by MORDELL several years ago.[9]

WEIL's theorem suggests a new approach to attack FERMAT's theorem and more generally the theory of algebraic diophantine equations in two unknowns. However, the proof of the conjecture that any such equation of genus larger than 1 has only finite many solutions in rational numbers will require to overcome substantial difficulties. On the contrary, combining WEIL's ideas with the considerations which have resulted in the proof of THUE's theorem and moreover, in the first part of this exposition, to the proof of the transcendence of the BESSEL function $J_0(\xi)$ for algebraic $\xi \neq 0$, one can deduce a result which shows the exceptional position of the linear and of the indefinite quadratic equations in the theory of algebraic diophantine equations with two rational integer unknowns.[10]

Let $L(x, y) = ax + by$ be a homogeneous linear form with rational integer coefficients a and b. If c denotes a rational integer, which is divisible by the

[9]FOOTNOTE BY THE EDITORS: Here it is alluded to Mordell's contribution (elliptic curves) to the Mordell-Weil theorem. A few lines below Mordell's conjecture is alluded at, without attribution. This was proved in 1982 by G. Faltings; for more on this, see the article by the Editors in this volume.

[10]FOOTNOTE BY THE EDITORS: We point out that the theorem on integral points proved in this part of the paper can be restated as a result on irrationality of the G-functions representing an algebraic function on the curve. This represents a further link between the two parts of the paper, which seems to be confirmed by Siegel at this point.

greatest common divisor of a and b, then by BACHET the equation $L(x, y) = c$ has infinitely many solutions in rational integers x, y.

Let $Q(x, y) = ax^2 + bxy + cy^2$ be an indefinite quadratic form with rational integer coefficients a, b, c whose discriminant $b^2 - 4ac$ is not a perfect square. Then by LAGRANGE, for every rational integer number $d \neq 0$ representable by Q the equation $Q(x, y) = d$ has infinitely many solutions in rational integers x, y.

From these two types of binary diophantine equations with infinitely many integer solutions one easily gets generalisations. Let $A(u, v)$ and $B(u, v)$ be two arbitrary homogeneous polynomials of degree n with rational integer coefficients and not both proportional to L^n. Eliminating $u : v$ from the equations

$$x = \frac{A(u, v)}{L^n(u, v)}, \quad y = \frac{B(u, v)}{L^n(u, v)}$$

one obtains an algebraic equation involving x and y. This equation certainly has infinitely many solutions in rational integers x, y provided that the greatest common divisor of the coefficients a and b in L raised to the nth power divides the coefficients of A and B. Moreover, let $C(u, v)$ and $D(u, v)$ be two homogeneous integer polynomials of degree $2n$, not both proportional to Q^n, then the equations

$$x = \frac{C(u, v)}{Q^n(u, v)}, \quad y = \frac{D(u, v)}{Q^n(u, v)}$$

also define an algebraic equation involving x and y which has infinitely many integer solutions if one requires the existence of a number, representable by Q, whose nth power is contained in all coefficients of C and D.

These two famous types of diophantine equations with infinitely many integral solutions are easily deduced from the classical equations $L = c$ and $Q = d$. In this exposition it shall now be proved that in this way already *all* binary diophantine equations with infinitely many rational integer solutions are given. The assertion therefore is:

I. *Suppose that the algebraic equation $f(x, y) = 0$ is not solvable identically in a parameter t by setting $x = A : L^n, y = B : L^n$ or $x = C : Q^n, y = D : Q^n$, where A, B, C, D are integer polynomials in t, L is a linear, Q is an indefinite quadratic polynomial in t. Then the equation has only finitely many solutions in rational integers.*

The exceptions obviously require that the genus of the equation has the value $p = 0$. It will first be shown that by MAILLET and by a certain argument the statement is true for $p = 0$. The real difficulty provides the case $p > 0$, *i.e.* the proof of the following statement:

An algebraic equation, whose genus is positive, has only finitely many solutions in rational integer numbers.

Proving the analogous theorem for algebraic number fields requires the same effort. Thereby it is advantageous for the formulation of the theorem to further broaden the assumptions. From now onwards, for solving the equation

$f(x, y) = 0$ also fractional numbers x and y from the algebraic number field $\mathfrak{K}$ will be allowed, but only such fractions for which multiples cx, cy for a *fixed* integer c are again integers. Such numbers will be called *quasi-integral*[11]. Then the following generalisation of the previously stated theorem holds:

II. *For the existence of infinitely many quasi-integral solutions in an algebraic number field for the irreducible equation $f(x, y) = 0$ it is necessary and sufficient that the equation $f = 0$ can be identically parameterized in t by two rational functions $x = P(t), y = Q(t)$; the functions P and Q shall be given by*

$$P(t) = a_n t^n + a_{n-1} t^{n-1} + \cdots + a_{-n} t^{-n} \tag{75}$$

$$Q(t) = b_n t^n + b_{n-1} t^{n-1} + \cdots + b_{-n} t^{-n} \tag{76}$$

and are assumed not to be both constant.

This can also be expressed in a different way. One can obviously assume that t is a rational function in x and y with algebraic coefficients. In fact, it follows from the parametrization $x = P, y = Q$ that f must have genus 0. Now express x and y as rational functions of a parameter τ which in turn is a rational function of x and y; then τ is a rational function in t. Since τ is determined only up to a linear transformation, one can assume that for $t = 0$ also $\tau = 0$ and for $t = \infty$ either $\tau = 0$ or $\tau = \infty$. But P and Q have, as functions of t, no other poles except at most the places 0 and ∞, hence the same is true for the dependence on τ. Therefore equations of the form as in (75) and (76) also hold with τ instead of t and one can therefore assume that t is already the uniformizing parameter. The coefficients $a_n, \ldots, b_{-n}$ in (75) and (76) are determined by substituting $x = P(t), y = Q(t)$ in $f = 0$; hence they can be chosen as algebraic numbers. The same is true for the coefficients of the rational function $R(x, y)$, which expresses t through x and y. If now Q is not constant, the substitution

$$x = P(t) + u, \quad y = Q(t)$$

transforms the equation $f(x, y) = 0$ into $u = 0$. But if Q is constant, hence P is not constant, the same result is obtained through the substitution

$$x = P(t) + u, \quad y = Q(t) - u.$$

In case P and Q do not contain any negative powers of t, the birational substitutions on the curve

$$x = P(t) + u, \quad y = Q(t); \qquad u = 0, \quad t = R(x, y)$$

$$x = P(t) + u, \quad y = Q(t) - u; \qquad u = 0, \quad t = R(x, y)$$

[11] Footnote by the Editors: In the original paper Siegel uses the word *ganzartig*.

respectively, transforms all the points of the curve $f = 0$, whose coordinates x, y are quasi-integral numbers in an algebraic number field, into points of the curve $u = 0$, whose coordinates u, t are also quasi-integral numbers in an algebraic number field; and viceversa. In case P and Q do not contain any positive powers of t, then on replacing t by $\frac{1}{t}$ one reduces to the case just discussed. In the final case that at least one of the functions P, Q contains both positive and negative powers of t, one replaces the negative powers $\frac{1}{t}$ with u. Then the birational substitution

$$\begin{aligned}
x &= a_n t^n + \cdots + a_1 t + a_0 + a_{-1} u + \cdots + a_{-n} u^n \\
y &= b_n t^n + \cdots + b_1 t + a_0 + b_{-1} u + \cdots + b_{-n} u^n \\
u &= 1 : R(x, y), \quad t = R(x, y)
\end{aligned}$$

again transforms the points of the curve $f = 0$, whose coordinates x, y are quasi-integral numbers in an algebraic number field, in analogous points u, t of the curve $ut = 1$; and viceversa. Therefore the theorem can be expressed as follows:

For $f(x, y) = 0$ to have infinitely many quasi-integral solutions in an algebraic number field it is necessary and sufficient that the equation $f = 0$ can be transformed either into $u = 0$ or into $ut = 1$ by a birational transformation which links all pairs x, y and u, t to each other.

Apart from these birational transformations only the trivial equations $u = 0$ and $ut = 1$ have infinitely many solutions in quasi-integral numbers in an algebraic number field.

Taking into account that t assumes every value only once on the RIEMANN surface of the algebraic structure defined by $f = 0$, it is clear that the two cases $u = 0$ and $ut = 1$ can be distinguished as follows: In the first case the quantity $|x| + |y|$ becomes ∞ at only one point on the RIEMANN surface, in the second case in exactly two points; and this is also sufficient for the existence of (75) and (76). Therefore one has the following very simple statement:

For the equation $f(x, y) = 0$ to have infinitely many quasi-integral solutions in an algebraic number field it is necessary and sufficient that the corresponding RIEMANN *surface has genus* 0 *and contains at most two poles of the function* $|x| + |y|$.

The two cases $u = 0$ and $ut = 1$ are essentially different from each other. This can be seen from the number of solutions that they have below a given bound. Namely, if one only considers the quasi-integral solutions x, y from an algebraic number field for which the absolute values of all conjugates of x and y are below the bound M, then, as easily follows from the above, the exact order of the number of solutions is M^{χ} in the first case and $(\log M)^{\lambda}$ in the second case, where χ and λ are suitable positive numbers only depending on f and on the field.

In the proof implicitly a finite upper bound for the number of solutions is obtained valid for every equation $f = 0$ not belonging to one of the exceptional cases. A bound for the solutions themselves however is not obtained; the problem of finding the finitely many solutions therefore remains open.[12]

The method of the proof also gives without any modifications an analogous result for space curves: A system of $n-1$ independent algebraic equations in n unknowns $x_1, \ldots, x_n$ with algebraic coefficients has infinitely many quasi-integral solutions in an algebraic number field if and only if $x_1, \ldots, x_n$ can be parameterized by polynomials in t and t^{-1}.

Another generalisation refers to the case $p > 0$. The finiteness of the number of solutions of $f(x, y) = 0$ will actually be shown under the weaker hypothesis that only one of the unknowns is quasi-integral. Therefore if $f = 0$ has infinitely many solutions x, y in an algebraic number field, then for these neither the norm of the denominator of x nor the norm of the denominator of y is bounded.

Finally, an application to HILBERT's Irreducibility Theorem can be mentioned. Let $P(w, x, y, \ldots)$ denote a polynomial in the variables $w, x, y, \ldots$, having integer coefficients in an algebraic number field $\mathfrak{K}$. If one considers P as function of $x, y, \ldots$ alone, then the coefficients are polynomials in w; they shall be denoted by $a_1(w), a_2(w), \ldots$ in some order. For a fixed numerical value of w the polynomial P is assumed to be reducible over $\mathfrak{K}$. Then $P = QR$, where Q and R are polynomials in $x, y, \ldots$, whose coefficients $b_1, b_2, \ldots$ and $c_1, c_2, \ldots$ belong to $\mathfrak{K}$. Apart from a constant factor, there are for fixed w only finitely many possibilities for Q and R, *i.e.* the products $b_k c_l = d_{kl}$ $(k = 1, 2, \ldots; l = 1, 2, \ldots)$ are determined by w in a finite range of alternatives. By a theorem due to GAUSS and KRONECKER, these products are integers, when w is an integer. The matrix (d_{kl}) has rank 1. By comparison of coefficients the numbers $a_1(w), a_2(w), \ldots$ are given as sums of certain of the d_{kl}. This leads to a system of algebraic equations in w and in the d_{kl}. Therefore for an arbitrary w in the space of the w and d_{kl}, either a system of a finite number of points or a space curve is obtained. A necessary and sufficient condition for the existence of infinitely many points on this curve, whose coordinates w and d_{kl} $(k = 1, 2, \ldots; l = 1, 2, \ldots)$ are quasi-integral in an algebraic number field, has been given above. It follows that:

For the existence of an algebraic number field $\mathfrak{K}$, in which the polynomial P is reducible for infinitely many quasi-integral values of $\mathfrak{K}$, it is necessary and sufficient that P, after a suitable substitution of the form

$$w = \alpha_n t^n + \alpha_{n-1} t^{n-1} + \cdots + \alpha_{-n} t^{-n},$$

is identically reducible in t.

As above, in this exceptional case one obtains a precise estimate of the density of those infinitely many quasi-integral numbers w that make P reducible.[13]

[12]FOOTNOTE BY THE EDITORS: We refer to the article on integral points, which we have included after the translation, for recent developments on this effectivity issues, still open in the general case.

[13]FOOTNOTE BY THE EDITORS: Similar estimates are possible using the result in [E. Bombieri,

§1. Equations of genus 0.

If the equation $f(x, y) = 0$ has genus 0, the following uniformization is possible[14]

$$x = \frac{\phi(u, v)}{\chi(u, v)}, \quad y = \frac{\psi(u, v)}{\chi(u, v)}, \quad \frac{u}{v} = \frac{A(x, y)}{B(x, y)} \tag{77}$$

where ϕ, ψ, χ denote homogeneous polynomials of the same degree and A, B are polynomials. The coefficients of f are assumed to belong to the algebraic number field $\mathfrak{K}$. It is further assumed that the curve $f = 0$ has a regular point, whose coordinates $x = \xi, y = \eta$ belong to $\mathfrak{K}$. If at this point $\frac{dy}{dx} = \infty$, then in the sequel the role of x and y are changed.

By (77), the function $\frac{u}{v} = t$ on the curve is determined up to a linear transformation. Since t assumes every value on the curve precisely once, it follows that in particular the value of $\frac{dt}{dx}$ at the regular point ξ, η is different from 0. One can therefore assume that in this point t has the value 0, while $\frac{dt}{dx}$ has the value 1. Finally, it can also be assumed that in the point ξ, η the value $\frac{d^2t}{dx^2} = 0$, because $\frac{t}{1+ct}$ has there the second derivative $\frac{d^2t}{dx^2} - 2c$. The three mentioned properties completely determine t. It will now be shown that the coefficients of the polynomials ϕ, ψ, χ, A, B may be chosen to be integers of $\mathfrak{K}$. In any case, these coefficients can be chosen as algebraic numbers; let these belong to a field $\mathfrak{L}$, which contains $\mathfrak{K}$. One then replaces in (77) all coefficients with the values that correspond to them in a field $\mathfrak{L}'$ conjugate to $\mathfrak{L}$ with respect to $\mathfrak{K}$. Since one obtains a uniformization again, the new function t evolves from the old one through a linear transformation, and since the values $0, 1, 0$ remain invariant when passaging from $\mathfrak{L}$ to $\mathfrak{L}'$, the same is true for the function t. Taking the arithmetic mean of the three equations (77) over all the conjugate equations with respect to $\mathfrak{K}$, then all coefficients of the right-hand side will be numbers of $\mathfrak{K}$.

It can be assumed that the three polynomials ϕ, ψ, χ are coprime. By the Euclidean algorithm it therefore holds that

$$P_1\phi + Q_1\psi + R_1\chi = \lambda_1 u^h \tag{78}$$

$$P_2\phi + Q_2\psi + R_2\chi = \lambda_2 v^h, \tag{79}$$

where $P_1, Q_1, R_1, P_2, Q_2, R_2$ denote homogeneous polynomials in u, v with integer coefficients from $\mathfrak{K}$, h is a natural number and λ_1, λ_2 are non-zero integers from $\mathfrak{K}$. If x and y are numbers from $\mathfrak{K}$, then according to (77) the values

J. Pila, The number of integral points on arcs and ovals, Duke Math. J. **59** (1989), 337-357] and the subsequent ones by D.R. Heath-Brown as e.g. [D. R. Heath-Brown, The density of rational points on curves and surfaces, Ann. of Math. (2) **155** (2002), 553–595].

[14] Footnote by the Editors: A plane curve is tacitly identified with a nonsingular model (which could not be presentable as a plane curve).

of u and v can be chosen to be integers of $\mathfrak{K}$. Let $(u, v) = \delta$ be a greatest common divisor of u and v. If x and y are integer numbers in $\mathfrak{K}$, then by (77) the number $\chi\left(\frac{u}{\delta}, \frac{v}{\delta}\right)$ divides the numbers $\phi\left(\frac{u}{\delta}, \frac{v}{\delta}\right)$ and $\psi\left(\frac{u}{\delta}, \frac{v}{\delta}\right)$, and so, using (78) and (79), it also divides the numbers $\lambda_1 (u/\delta)^h$ and $\lambda_2 (v/\delta)^h$, hence also $\lambda_1\lambda_2$. Therefore one has an equality

$$\chi\left(\frac{u}{\delta}, \frac{v}{\delta}\right) = \gamma, \tag{80}$$

where γ is a divisor of the fixed number $\lambda_1\lambda_2$.

Furthermore, let $\mathfrak{K}$ be the field of rational numbers. Then γ is one of finitely many rational integers; χ is a homogeneous polynomial with rational integer coefficients; $\frac{u}{\delta}$ and $\frac{v}{\delta}$ are rational integer numbers. By THUE's theorem the equation (80) can have infinitely many solutions $\frac{u}{\delta}, \frac{v}{\delta}$ only if $\chi(u, v)$ is a power of either a linear or of an indefinite quadratic polynomial. On the other hand, different values of $\frac{u}{v}$ correspond to different pairs x, y. Therefore assertion I. in the case $p = 0$ is proved.

From now on, let $\mathfrak{K}$ be an arbitrary number field. If among the linear factors of $\chi(u, v)$ at most two are different from each other, *i.e.* if x and y become infinitely large for at most two values of $u : v$, then by a linear transformation of $u : v$ it is possible to transfer these values to 0 and ∞ and thus expressions of the form (75) and (76) are obtained for x and y. In case $\chi(u, v)$ contains more than two different linear factors $u - \alpha_1 v, u - \alpha_2 v, u - \alpha_3 v, \ldots$, then one can assume without loss of generality that $\alpha_1, \alpha_2, \alpha_3, \ldots$ are integers of $\mathfrak{K}$, because otherwise it would suffice to choose v equal to a suitable multiple of a new variable and to suitably extend the field $\mathfrak{K}$. But then the numbers

$$\frac{u}{\delta} - \alpha_1\frac{v}{\delta}, \quad \frac{u}{\delta} - \alpha_2\frac{v}{\delta}, \quad \frac{u}{\delta} - \alpha_3\frac{v}{\delta}$$

are all divisors of the fixed number $\lambda_1\lambda_2$. In the equation[15]

$$\frac{\alpha_3 - \alpha_2}{\alpha_1 - \alpha_2}\,\frac{u - \alpha_1 v}{u - \alpha_3 v} + \frac{\alpha_3 - \alpha_1}{\alpha_2 - \alpha_1}\,\frac{u - \alpha_2 v}{u - \alpha_3 v} = 1 \tag{81}$$

only finitely many non-associate numbers for $\frac{u-\alpha_1 v}{u-\alpha_3 v}$ and $\frac{u-\alpha_2 v}{u-\alpha_3 v}$ are possible. Let n be a natural number. By a result of DIRICHLET the group of the nth powers of the units of $\mathfrak{K}$ has finite index in the group of all units of $\mathfrak{K}$. Therefore

$$\frac{u - \alpha_1 v}{u - \alpha_3 v} = \gamma_1 \epsilon_1^n, \qquad \frac{u - \alpha_2 v}{u - \alpha_3 v} = \gamma_2 \epsilon_2^n, \tag{82}$$

where for the numbers γ_1 and γ_2 only finitely many values are possible and ϵ_1, ϵ_2 are units, hence integers of $\mathfrak{K}$. For each of the finitely many pairs for

[15] FOOTNOTE BY THE EDITORS: This is the celebrated *unit equation*, whose importance is stressed here for one of the first times.

γ_1, γ_2 one has, according to (81), that

$$\frac{\alpha_3 - \alpha_2}{\alpha_1 - \alpha_2}\gamma_1\epsilon_1^n + \frac{\alpha_3 - \alpha_1}{\alpha_2 - \alpha_1}\gamma_2\epsilon_2^n = 1. \tag{83}$$

Using a generalisation of THUE's theorem, which will be proved below in a more general context but has already been known before, the equation (83) has only finitely many solutions in integer numbers ϵ_1, ϵ_2 from $\mathfrak{K}$ provided that the degree n is larger than a bound depending only on the field. Hence, by (82), one has only finitely many values $\frac{u}{v}$ and, using (77), only finitely many solutions of the equation $f(x, y) = 0$ in integers x, y from $\mathfrak{K}$.

Therefore the necessity of the condition in assertion II. for the existence of infinitely many integer solutions from $\mathfrak{K}$ in the case $p=0$ is proved. The quasi-integralness can immediately be reduced to the integralness by substituting x, y with $\frac{x}{c}, \frac{y}{c}$ for a suitable fixed natural number c. The fact that the condition in assertion II. is also sufficient for the existence of infinitely many quasi-integral solutions of the equation $f=0$ in a suitable algebraic number field is trivial; one only needs, for instance, to replace in (75) and (76) t with all the powers of $1+\sqrt{2}$ and then obtains infinitely many quasi-integral numbers x, y all belonging to the same algebraic number field. Hence, assertion II. is proved in the case $p=0$.

The proof is based on the reduction of the given equation to (83), where then THUE's methods can be used. Also in the case $p > 0$ the equation $f = 0$, to be solved in integers, will be transformed into another diophantine equation, whose integer solutions give good approximations to a fixed algebraic number; and the considerations sketched in the introduction of this exposition will then show that such a good approximation is possible in just a finite number of times. The analogy between algebraic and arithmetic divisibility, which is expressed in (78) and (79) and has been used in the proof above, can be transformed to algebraic functions, as noticed by A. WEIL; the parallelism between ideals in function fields and ideals in number fields that he discovered shall be presented in §2 with some minor modifications.[16]

§2. Ideals in function fields and number fields.

Assume that the variables x and y are related by an equation $f(x, y) = 0$ of genus p. Replace x, y with $\frac{x}{z}, \frac{y}{z}$ and consider the ring $\mathfrak{R}$ of those homogeneous rational functions in x, y, z that do not become infinite for any finite x, y, z and have algebraic coefficients. Suppose that x, y, z assume only coprime integral algebraic values; then the values of the functions in $\mathfrak{R}$ are algebraic numbers too.

[16] FOOTNOTE BY THE EDITORS: The link between arithmetical properties in number fields and function fields is quite relevant throughout in this paper. For later developments, see e.g. [E. Bombieri, On Weil's "théorème de décomposition", Amer. J. Math. **105** (1983), 295–308], where principles appearing also in this present setting are discussed.

Now let ϕ, ψ, χ be three arbitrary forms from $\mathfrak{R}$ having the same degree n and such that the quotient $\psi : \chi$ is not constant. Then $\phi : \chi$ is an algebraic function of $\psi : \chi$ and an irreducible equation

$$\phi^m + R_1(\psi, \chi)\phi^{m-1} + \cdots + R_m(\psi, \chi) = 0 \tag{84}$$

holds, where the coefficients $R_1(\psi, \chi), \ldots, R_m(\psi, \chi)$ are homogeneous rational functions of ψ, χ of degree $1, \ldots, m$ having algebraic coefficients. It will be shown that $R_1, \ldots, R_m$ are polynomials if at no point on the curve both forms ψ and χ vanish simultaneously of higher order than the form ϕ. If one of the functions $R_1, \ldots, R_m$ had a linear factor $a\psi + b\chi$ in its denominator, then for *arbitrary* constants α, β the function $\phi : (\alpha\psi + \beta\chi)$ would become infinite for $a\psi + b\chi = 0$; then at the same time $\psi = 0, \chi = 0$; fixing now $\alpha = 0$ or $\beta = 0$ one derives a contradiction to the assumption about the vanishing of ϕ. Now let c be such an integer $\neq 0$, so that the coefficients of the polynomials $cR_1, \ldots, cR_m$ are integers. From (84) it follows:

If the two forms ψ and χ do not vanish at any point simultaneously with higher order than the form ϕ, then a constant $c \neq 0$ exists so that every numerical common divisor of ψ and χ divides $c\phi$ too.

Choosing particularly $\psi = (a_1x + b_1y + c_1z)^n, \chi = (a_2x + b_2y + c_2z)^n$, where the coefficients $a_1, \ldots, c_2$ are integers such that ψ and χ do not vanish simultaneously, then for every form ϕ of $\mathfrak{R}$ a constant $c \neq 0$ exists so that $c\phi$ is an integer.

More generally, let $\phi_1, \ldots, \phi_k$ be arbitrary forms of $\mathfrak{R}$, not necessarily having the same degree, and let Φ be a form such that the functions $\phi_1, \ldots, \phi_k$ do not vanish at any point simultaneously of higher order than the form Φ itself. If $\phi_1, \ldots, \phi_k$ are all proportional, then $\Phi : \phi_1, \ldots, \Phi : \phi_k$ are again functions of $\mathfrak{R}$; hence a constant $c \neq 0$ exists so that $c\Phi : \phi_1, \ldots, c\Phi : \phi_k$ are integers and so that every numerical common divisor of $\phi_1, \ldots, \phi_k$ also divides $c\Phi$. But if $\phi_1, \ldots, \phi_k$ are not all proportional, then one determines forms $a_1, \ldots, a_k$ and $b_1, \ldots, b_k$ so that

$$a_1\phi_1 + \cdots + a_k\phi_k = \psi, \quad b_1\phi_1 + \cdots + b_k\phi_k = \chi$$

are both homogeneous and of a degree that is by a non-negative number r bigger than the degree of Φ; furthermore ψ and χ should not vanish at any point simultaneously with higher order than the form Φ. If now ϕ is consecutively chosen to be one of the three functions $x^r\Phi, y^r\Phi, z^r\Phi$, then from the above it follows that a constant $c \neq 0$ exists for which every numerical common divisor of $\phi_1, \ldots, \phi_k$ divides $c\Phi$ as well. Assume that the forms $\phi_1, \ldots, \phi_k$ have the same zeros in common as the forms $\psi_1, \ldots \psi_l$, also with the same multiplicities, then the greatest common divisor of the numbers $\phi_1, \ldots, \phi_k$ and $\psi_1, \ldots \psi_l$, respectively, differ only by a factor whose integer numerator and denominator divide a fixed number, which is independent of the point on the curve. Such a factor will be called *unitary*. Therefore the greatest common divisor of the numbers $\phi_1, \ldots, \phi_k$ is already determined up to a unitary factor by the common zeros of the functions $\phi_1, \ldots, \phi_k$.

The point on the curve with homogeneous coordinates x, y, z will be denoted by $\mathfrak{p}$. If $\mathfrak{p} = \mathfrak{p}_0$ is a fixed point, then one constructs two functions ϕ_1 and ϕ_2 that only have the root $\mathfrak{p}_0$ in common. The greatest common divisor of the numbers ϕ_1 and ϕ_2 is then a function of $\mathfrak{p}$ and will be denoted by $\omega(\mathfrak{p}, \mathfrak{p}_0)$; this number still depends on the choice of the forms ϕ_1, ϕ_2, but for fixed $\mathfrak{p}_0$ only up to a unitary factor. Observe now that from the kl products $\phi_1\psi_1, \ldots, \phi_k\psi_l$ it is possible to build through linear combinations two forms, whose common zeros are exactly formed by the common zeros of $\phi_1, \ldots, \phi_k$ and the common zeros of $\psi_1, \ldots \psi_l$. Therefore the greatest common divisor of the numbers $\phi, \psi, \ldots$ is up to a unitary factor equal to $\omega(\mathfrak{p}, \mathfrak{p}_1) \cdots \omega(\mathfrak{p}, \mathfrak{p}_r)$, where $\mathfrak{p}_1, \ldots, \mathfrak{p}_r$ denote the common zeros of the forms $\phi, \psi, \ldots$.

For what follows, an estimate of the arithmetic norm of $\omega(\mathfrak{p}, \mathfrak{p}_0)$ as a function of $\mathfrak{p}$ is needed. Let the order of the curve $f = 0$ be h. The form

$$L = ax + by + cz$$

vanishes at h points. By ABEL's theorem there is a function in the algebraic function field generated by $\frac{x}{z}$ and $\frac{y}{z}$, which has no other poles apart at most of the hm zeros of L^m and which vanishes at $hm - p$ prescribed points; this function shall be called ϕ. Then $\phi L^m = \psi$ is a form of $\mathfrak{R}$ of degree exactly m and vanishing at $hm - p$ prescribed points. Choose the point $\mathfrak{p}_0$ counted $hm - p$-times as such points. The remaining p roots of ψ will be called $\mathfrak{q}_1, \ldots, \mathfrak{q}_p$; they depend on $\mathfrak{p}_0$ and on the coefficients of L. The coefficients of ψ are algebraic numbers and, up to a unitary factor, ψ is equal to $(\omega(\mathfrak{p}, \mathfrak{p}_0))^{hm-p}\omega(\mathfrak{p}, \mathfrak{q}_1) \cdots \omega(\mathfrak{p}, \mathfrak{q}_p)$. On the other hand, ψ is finite for all x, y, z, hence it is bounded for $|x| + |y| + |z| = 1$; and from the homogeneity the inequality

$$|\psi| < c(|x| + |y| + |z|)^m \tag{85}$$

follows at once, where c depends on $\mathfrak{p}_0$ and m but not on x, y, z.

Let the coordinates x, y, z of $\mathfrak{p}$ belong to the field $\mathfrak{K}$ of the coefficients of f. The field that is obtained by adjoining the coordinates of $\mathfrak{p}_0$ to $\mathfrak{K}$ will be denoted by $\mathfrak{K}(\mathfrak{p}_0)$. If the coefficients a, b, c of L belong to $\mathfrak{K}$, then the coefficients of ψ can be chosen from $\mathfrak{K}(\mathfrak{p}_0)$. The inequality (85) then also holds in the conjugates fields of $\mathfrak{K}(\mathfrak{p}_0)$ with respect to $\mathfrak{K}$. If r denotes the relative degree of $\mathfrak{K}(\mathfrak{p}_0)$ and $\mathfrak{p}_0, \mathfrak{p}_0', \ldots, \mathfrak{p}_0^{(r-1)}$ the conjugates of $\mathfrak{p}_0$ with respect to $\mathfrak{K}$ and analogously $\mathfrak{q}_1, \mathfrak{q}_1', \ldots, \mathfrak{q}_1^{(r-1)}$ the conjugates of $\mathfrak{q}_1, \ldots$, then the product

$$(\omega(\mathfrak{p},\mathfrak{p}_0)\omega(\mathfrak{p},\mathfrak{p}_0') \cdots)^{hm-p}(\omega(\mathfrak{p},\mathfrak{q}_1)\omega(\mathfrak{p},\mathfrak{q}_1') \cdots)(\omega(\mathfrak{p}, \mathfrak{q}_2)\omega(\mathfrak{p},\mathfrak{q}_2') \cdots) \cdots \tag{86}$$

lies in $\mathfrak{K}$. Finally, take conjugates (85) over $\mathfrak{K}$ with respect to the field of rational numbers; then the norm of the expression (86) is less than

$$c_1 N(|x| + |y| + |z|)^{mr},$$

where c_1 depends on $\mathfrak{p}_0$ and m. A fortiori it holds

$$(87)\qquad N\left(\left|\omega(\mathfrak{p},\mathfrak{p}_0)\cdots\omega(\mathfrak{p},\mathfrak{p}_0^{(r-1)})\right|\right) < c_2 N(|x|+|y|+|z|)^{\frac{mr}{hm-p}},$$

with analogous meaning of c_2; here the symbol N means that the product is taken over all conjugate fields of $\mathfrak{K}$. In (87) the value of m can be arbitrarily large; consequently one has

$$(88)\qquad N\left(\left|\omega(\mathfrak{p},\mathfrak{p}_0)\cdots\omega(\mathfrak{p},\mathfrak{p}_0^{(r-1)})\right|\right) < c_3 N(|x|+|y|+|z|)^{\frac{r}{h}+\epsilon},$$

for every $\epsilon > 0$, where c_3 depends on ϵ and $\mathfrak{p}_0$.

Consider two forms χ_1, χ_2 of the same degree, with coefficients in $\mathfrak{K}$ and with $\mathfrak{p}_1, \mathfrak{p}_2, \ldots$ their common zeros. An arbitrary form of the family $\lambda_1\chi_1 + \lambda_2\chi_2$ then has in addition to $\mathfrak{p}_1, \ldots$ also certain other roots $\mathfrak{r}, \ldots$. Let λ_1, λ_2 be rational numbers. If one replaces the coefficients of χ_1 and χ_2 simultaneously by their conjugates, then the zeros $\mathfrak{r}', \ldots; \mathfrak{r}'', \ldots; \ldots$ shall replace the zeros $\mathfrak{r}, \ldots$; denote the totality of these zeros $\mathfrak{r}, \ldots; \mathfrak{r}', \ldots; \ldots$ by $\mathfrak{S}$. For each l a family of $l+1$ forms can be determined, say $\psi_0, \ldots, \psi_l$, so that no two of the corresponding root systems $\mathfrak{S}_0, \ldots, \mathfrak{S}_l$ have any elements in common; one just has to observe in the construction of such forms that λ_1, λ_2 can certainly be chosen so that $\lambda_1\chi_1 + \lambda_2\chi_2$ does not vanish at certain finitely many points in which χ_1 and χ_2 do not both vanish. If $\mathfrak{q}_1, \ldots, \mathfrak{q}_l$ are any l points, then $\mathfrak{q}_1$ appears in at most one of the systems $\mathfrak{S}_0, \ldots, \mathfrak{S}_l$, say in $\mathfrak{S}_0$ and then it does not appear in the systems $\mathfrak{S}_1, \ldots, \mathfrak{S}_l$; similarly $\mathfrak{q}_2$ appears in at most one of the systems $\mathfrak{S}_1, \ldots, \mathfrak{S}_l$, say in $\mathfrak{S}_1$ and then it does not appear in the systems $\mathfrak{S}_2, \ldots, \mathfrak{S}_l; \ldots$; consequently there is a system which does not contain any of the points $\mathfrak{q}_1, \ldots, \mathfrak{q}_l$.

Let Ψ be a function of the algebraic function field obtained by $\frac{x}{z}$ and $\frac{y}{z}$ with coefficients in $\mathfrak{R}$. Let its zeros be $\mathfrak{p}_1, \ldots, \mathfrak{p}_g$; let its poles be $\mathfrak{q}_1, \ldots, \mathfrak{q}_g$. Let l be the degree of the field $\mathfrak{K}$ and let $\mathfrak{p}$ be in $\mathfrak{K}$ as above. According to what has been shown above, one can choose among certain $l+1$ forms $\psi_0, \ldots, \psi_l$, which are determined by $\mathfrak{p}_1, \ldots, \mathfrak{p}_g$, a function ψ that vanishes at $\mathfrak{p}_1, \ldots, \mathfrak{p}_g$ but whose other roots $\mathfrak{r}, \ldots$ as well as the conjugates of these roots are not near $\mathfrak{p}$ and the conjugates of $\mathfrak{p}$; moreover, the coefficients of the form ψ are numbers from $\mathfrak{K}$. Then $\psi : \Phi = \chi$ is a form that vanishes at the points $\mathfrak{q}_1, \ldots, \mathfrak{q}_g, \mathfrak{r}, \ldots$, that becomes infinite nowhere, and that has the same degree γ as ψ. If now $\mathfrak{p}$ does not lie in a neighbourhood of the roots $\mathfrak{q}_1, \ldots, \mathfrak{q}_g, \mathfrak{r}, \ldots$ of χ, then analogously to (85) the estimate

$$|\chi| > c(|x|+|y|+|z|)^{\gamma}$$

holds, and also

$$c\,|\Phi|\,(|x|+|y|+|z|)^{\gamma} < |\psi|,$$

where $c > 0$ depends on the size of the forbidden neighbourhood for $\mathfrak{p}$. This inequality also holds in the conjugate fields, if one also assumes that $\mathfrak{p}$ is outside the neighbourhoods of the roots of the conjugate functions of χ. Thus,

$$N(|\Phi|)\,N(|x|+|y|+|z|)^{\gamma} < c_4 N\left(\left|\omega(\mathfrak{p},\mathfrak{p}_1)\cdots\omega(\mathfrak{p},\mathfrak{p}_g)\right|\,N\,|\omega(\mathfrak{p},\mathfrak{r})\cdots|\right).$$

For the factor $N\,(|\omega(\mathfrak{p},\mathfrak{r})\cdots|)$ one uses the estimate (88), where one pays attention to the fact that the form ψ has exactly $h\gamma$ roots, hence the number of zeros $\mathfrak{r},\ldots$ equals $h\gamma-g$; it becomes

$$N\,(|\omega(\mathfrak{p},\mathfrak{r})\cdots|) < c_5 N(|x|+|y|+|z|)^{\frac{h\gamma-g}{h}+\epsilon}$$

(89) $$N\,(|\Phi|) < c_6 N\left|\omega(\mathfrak{p},\mathfrak{p}_1)\cdots\omega(\mathfrak{p},\mathfrak{p}_g)\right| N(|x|+|y|+|z|)^{-\frac{g}{h}+\epsilon}.$$

This holds under the condition that the conjugates of $\mathfrak{p}$ do not lie near the poles $\mathfrak{q}_1,\ldots,\mathfrak{q}_g$ of Φ.

The inequality (89) provides the possibility to deduce from the assumption that a function of the algebraic function field obtained by $\frac{x}{z}$ and $\frac{y}{z}$ is integral for infinitely many values of $\frac{x}{z}$ and $\frac{y}{z}$ in $\mathfrak{K}$, an assertion about the approximation of a certain algebraic number. In fact, let F be such a function with coefficients in $\mathfrak{K}$ of order g, let its poles be $\mathfrak{p}_1,\ldots,\mathfrak{p}_g$, let its roots be $\mathfrak{q}_1,\ldots,\mathfrak{q}_g$. Let $\mathfrak{p}$ run through the points in which F is integral. Since there are two forms, which do not have any root in common, the first vanishing in $\mathfrak{p}_1,\ldots,\mathfrak{p}_g$ and the second in $\mathfrak{q}_1,\ldots,\mathfrak{q}_g$, it follows that $\omega(\mathfrak{p},\mathfrak{p}_1)\cdots\omega(\mathfrak{p},\mathfrak{p}_g)$ and $\omega(\mathfrak{p},\mathfrak{q}_1)\cdots\omega(\mathfrak{p},\mathfrak{q}_g)$ have only a unitary divisor in common. On the other hand, the quotient of these two numbers is up to a unitary factor equal to the integer F. Consequently, the expression $\omega(\mathfrak{p},\mathfrak{p}_1)\cdots\omega(\mathfrak{p},\mathfrak{p}_g)$ itself is unitary. Substituting F by one of the functions $F, F+1,\ldots,F+l$ if necessary, one can achieve that $\mathfrak{p}$ does not lie near to a root of F and the same holds true for all l conjugates of $\mathfrak{p}$ and F. Apply (89) with $\Phi = 1:F$; it follows

(90) $$N\,(|\Phi|) < c_7 N(|x|+|y|+|z|)^{-\frac{g}{h}+\epsilon}.$$

For the infinitely many $\mathfrak{p}$ that make F integral, there is a certain one of the l conjugates of Φ that is infinitely many times the smallest in absolute value; without loss of generality assume that this holds for Φ itself. The left-hand side of (90) is then not smaller than $|\Phi|^l$. Moreover, as is easily seen, the right-hand side converges to 0. In fact, if $\frac{x}{z},\frac{x'}{z'},\ldots$ are the l conjugates of $\frac{x}{z}$, then

$$(zt-x)(z't-x')\cdots = 0$$

is an equation of degree l with rational integer coefficients for $t=\frac{x}{z}$; and obviously these coefficients are in absolute value smaller than $c_8 N(|x|+|z|)$. Thus $c_8 N(|x|+|y|+|z|)$ is an upper bound for the absolute values of the rational integer coefficients of the equations, which are satisfied by $\frac{x}{z}$ and $\frac{y}{z}$. Because of the existence of infinitely many $\mathfrak{p}$ the value $N(|x|+|y|+|z|)$ becomes infinite.

By (90), the convergence of a subsequence of $\mathfrak{p}$ to a fixed root of Φ, say $\mathfrak{p}_1$, follows. Assume that its order is r. Let ϕ denote an arbitrary rational function of $\frac{x}{z}$ and $\frac{y}{z}$ that vanishes at $\mathfrak{p}_1$. Then $\phi^r : \Phi$ is bounded for $\mathfrak{p} \to \mathfrak{p}_1$. Therefore the following essential result is obtained:

Let $f\left(\frac{x}{z}, \frac{y}{z}\right) = 0$ be the equation of an algebraic curve of order h with coefficients from an algebraic number field $\mathfrak{K}$ of degree l. Assume that there is a function of order g in the algebraic function field obtained by $\frac{x}{z}$ and $\frac{y}{z}$ with coefficients in $\mathfrak{K}$, whose value is integral for infinitely many points on the curve $\frac{x}{z}, \frac{y}{z}$ belonging to $\mathfrak{K}$. Then a subsequence of these points on the curve converges to a pole of the function. Let r be the order of this pole. Then for any function ϕ of the function field vanishing at that pole the inequality

$$|\phi| < c_9 N(|x| + |y| + |z|)^{-\frac{g}{hlr}+\epsilon} \tag{91}$$

holds, provided that $\frac{x}{z}, \frac{y}{z}$ runs through the subsequence, the numbers x, y, z are chosen coprime and ϵ is an arbitrarily small positive number; the number c_9 depends on ϵ but not on x, y, z.

This is an approximation theorem for a zero of ϕ. For its application, it is important that it is possible to choose the number g sufficiently large. This is made possible through the branching theory for ABEL functions together with WEIL's theorem mentioned at the beginning.[17] Below curves of genus 1 will be treated, for which some difficulties of the general case do not appear.

§3. Equations of genus 1.

If the equation $f(x, y) = 0$ has genus 1, then a birational transformation

$$x = \phi(u, t), \quad y = \psi(u, t) \tag{92}$$

transforms it into the equation

$$t^2 = 4u^3 - g_2 u - g_3.$$

The coefficients of the rational functions ϕ and ψ, as well as the quantities g_2, g_3 are algebraic numbers; without loss of generality it can be assumed that they are contained in the field $\mathfrak{K}$, since otherwise it would just suffice to extend the field. If $\wp(s)$ denotes the WEIERSTRASS $\wp$-function obtained by the invariants g_2, g_3, then the ansatz

$$t = \wp(s), \quad u = \wp'(s) \tag{93}$$

uniformizes the curve $f = 0$. The field of elliptic functions corresponding to the invariants g_2, g_3 coincides with the algebraic function field obtained by x and y.

[17] FOOTNOTE BY THE EDITORS: The arguments above resemble the use of Weil functions in the context of modern height theory.

Let $w(s)$ be any non-constant function of this field and let r be its order, *i.e.* the number of times that it assumes any given value a in the period parallelogram. Let $\nu_1, \ldots, \nu_r$ be all the solutions of $w(s) = a$ in the period parallelogram written down with their multiplicities. If n denotes a natural number and c an arbitrary number, then the a-points of the elliptic function $w(ns+c)$ are exactly the points $s = \frac{1}{n}(\nu_k - c + \omega)$, where $k = 1, \ldots, r$ and ω is an arbitrary period. Within the period parallelogram for each a-point of order l of $w(s)$ exactly n^2 pairwise distinct a-points of order l for $w(ns+c)$ are obtained; thereby different a-points of $w(s)$ lead to different a-points of $w(ns+c)$. Consequently, the function $w(ns+c)$ has order $n^2 r = g$ and assumes no value more than r-times.

Consider all solutions in $\mathfrak{K}$ of $f(x, y) = 0$ in integral or fractional x, y. By the theorem of MORDELL and WEIL, the values of the integral of first kind s corresponding to these x, y form a module $\mathfrak{M}$ with finite basis. Let $s_1, \ldots, s_q$ be the elements of the basis. Then one obtains all the solutions in $\mathfrak{K}$ of $f = 0$ from (92) and (93) by putting therein

$$s = n_1 s_1 + \cdots + n_q s_q, \tag{94}$$

where $n_1, \ldots, n_q$ run through all rational integers. Let n be a natural number. By (94), every element of $\mathfrak{M}$ has the form

$$s = n\sigma + c, \tag{95}$$

where σ is an element of $\mathfrak{M}$ and c is one of finitely many elements of $\mathfrak{M}$; it suffices to restrict c to the values $n_1 s_1 + \cdots + n_q s_q$ with $n_k = 0, \ldots, n-1$ $(k = 1, \ldots, q)$.

It is now assumed that the equation $f = 0$ has infinitely many integer solutions in $\mathfrak{K}$. Among those, pick infinitely many for which the number c in (95) has a certain fixed value for all. Now identify x with the elliptic function $w(s) = w(n\sigma + c)$. If the solution ξ, η of $f = 0$ belongs to σ, then by the addition theorem x is a rational function of ξ, η with coefficients from $\mathfrak{K}$, whose order has the value $n^2 r = g$ and whose g poles have order at most r. The poles are determined by the solutions of an algebraic equation of degree g with coefficients in $\mathfrak{K}$. By the result of the previous section, one of the poles is approximated by infinitely many pairs of numbers ξ, η belonging to $\mathfrak{K}$. One replaces ξ, η by $\xi/\zeta, \eta/\zeta$ for coprime ξ, η, ζ.

If $\frac{\xi}{\zeta}$ converges to a finite limit point ρ, then ρ is at most of degree g with respect to $\mathfrak{K}$, and by (91) the inequality

$$\left| \frac{\xi}{\zeta} - \rho \right| < c_9 N(|\xi| + |\eta|)^{-\chi g + \epsilon} \tag{96}$$

holds; here $\chi = 1 : hlr$, where h is the order of the curve $f = 0$, l is the degree of the field $\mathfrak{K}$ and r is the degree of f in y; moreover, $g = n^2 r$ with n any natural number. As has already been mentioned before, $N(|\xi| + |\eta|)$ grows at least as fast as the largest of the absolute values of the rational integer coefficients of the equation of degree l for ξ/ζ, which will be denoted by $H(\xi/\zeta)$. By the previously mentioned generalisation of THUE's theorem, one has on the other hand

$$\left|\frac{\xi}{\zeta}-\rho\right| > c_{10}\left\{H\left(\frac{\xi}{\zeta}\right)\right\}^{-\lambda\sqrt{g}}, \tag{97}$$

where λ only depends on the degree of the field $\mathfrak{K}$ which contains the approximating number $\xi : \zeta$. However, for sufficiently large n, indeed for $n > \lambda hl\sqrt{r}$, (96) and (97) are surely satisfied only by finitely many $\xi : \zeta$.

If $\xi : \zeta$ converges to ∞, one has just to replace in the previous argument $\xi : \zeta$ and ρ by $\zeta : \xi$ and 0, and then comes to the same result.

In order to extend the proof to the general case $p \geq 1$, some auxiliary results concerning the branching of ABEL functions will first be needed.

§4. Auxiliary means from the theory of ABEL functions.

Let $\mathcal{R}$ be a RIEMANN surface of genus $p \geq 1$. Let it be canonically sliced by a system of p pairs of return slices $\mathcal{A}_l$, $\mathcal{B}_l$ $(l = 1, \ldots, p)$. Let $\mathfrak{p}$ be a variable point on the surface. As in the introduction, denote by $w_1(\mathfrak{p}), \ldots, w_p(\mathfrak{p})$ a system of normal integrals of the first kind. The period of w_k associated to $\mathcal{A}_l$ shall be denoted by e_{kl}, *i.e.* it is $= 0$ or $= 1$ depending on whether $k \neq l$ or $k = l$; the period of w_k associated to $\mathcal{B}_l$ shall be denoted by τ_{kl}. On the unsliced surface $w_1, \ldots, w_p$ are determined only up to the periods $\Omega_1, \ldots, \Omega_p$, where

$$\Omega_k = \sum_{l=1}^{p}(g_l e_{kl} + h_l \tau_{kl}) \qquad (k = 1, \ldots, p) \tag{98}$$

with $g_1, \ldots, g_p, h_1, \ldots, h_p$ rational integers. Two systems $a_1, \ldots, a_p$; $b_1, \ldots, b_p$ each consisting of p numbers, for which the differences $a_1 - b_1, \ldots, a_p - b_p$ are equal to a system of simultaneous periods $\Omega_1, \ldots, \Omega_p$ will be called congruent.

Let

$$\theta(s) = \theta(s_1, \ldots, s_p) = \sum_{n_1=-\infty}^{+\infty} \cdots \sum_{n_p=-\infty}^{+\infty} e^{\pi i \sum_{k,l}\tau_{kl}n_k n_l + 2\pi i \sum_k n_k s_k}$$

be the RIEMANN theta-function in the independent variables $s_1, \ldots, s_p$. There exist $p + 1$ fixed points $\mathfrak{a}, \mathfrak{a}_1, \ldots, \mathfrak{a}_p$ such that the inverse problem

$$\sum_{l=1}^{p}\{w_k(\mathfrak{p}_l) - w_k(\mathfrak{a}_l)\} \equiv s_k \qquad (k = 1, \ldots, p) \tag{99}$$

can be solved by a set of points $\mathfrak{p}_1, \ldots, \mathfrak{p}_p$ if and only if the theta-function with arguments $s_k - w_k(\mathfrak{p}) + w_k(\mathfrak{a})$, that is $\theta(s_k - w_k(\mathfrak{p}) + w_k(\mathfrak{a}))$, does not vanish identically in $\mathfrak{p}$. Denote the set of points $\mathfrak{p}_1, \ldots, \mathfrak{p}_p$ by $\mathfrak{P}$. Let $\pi_1, \ldots, \pi_p$ be any other set of points Π; analogously to (99) one sets

$$\text{(100)} \qquad \sum_{l=1}^{p} \{w_k(\pi_l) - w_k(\mathfrak{a}_l)\} \equiv \sigma_k. \qquad (k = 1, \ldots, p)$$

Furthermore, let n be a natural number and $c_1, \ldots, c_p$ be arbitrary constants; then the condition

$$\text{(101)} \qquad s_k \equiv n\sigma_k + c_k \qquad (k = 1, \ldots, p)$$

gives a relation between $\mathfrak{P}$ and Π. (101) yields

$$\text{(102)} \qquad \sigma_k \equiv \frac{1}{n}(s_k - c_k) + \frac{1}{n}\Omega_k,$$

where in the defining equation (98) of Ω_k one substitutes independently of each other $g_1, \ldots, g_p$; $h_1, \ldots, h_p$ by the integers $0, \ldots, n-1$; these lead to all n^{2p} incongruent systems $\frac{1}{n}\Omega_1, \ldots, \frac{1}{n}\Omega_p$. By (102) it follows that from the single system $s_1, \ldots, s_p$ one gets n^{2p} incongruent systems $\sigma_1, \ldots, \sigma_p$.

Consider now the special set of points $\mathfrak{P}_0$, which consists of the point $\mathfrak{p}$ counted p-times. The associated s_k has

$$s_k \equiv p w_k(\mathfrak{p}) - \sum_{l=1}^{p} w_k(\mathfrak{a}_l); \qquad (k = 1, \ldots, p)$$

and by (102) one gets

$$\text{(103)} \qquad \sigma_k \equiv \frac{p}{n} w_k(\mathfrak{p}) + b_k$$

with

$$b_k = \frac{1}{n}\left(-\sum_{l=1}^{p} w_k(\mathfrak{a}_l) - c_k + \Omega_k\right).$$

We shall now show that for all sufficiently large n there exists exactly one set of points Π which satisfies (100) and (103) if one ignores finitely many exceptional points $\mathfrak{p}$. For this purpose one has to show that the function

$$\text{(104)} \qquad \theta\left(\frac{p}{n} w(\mathfrak{p}) + b - w(\mathfrak{q}) + w(\mathfrak{a})\right)$$

vanishes identically in $\mathfrak{q}$ for at most finitely many $\mathfrak{p}$. If this were true for infinitely many $\mathfrak{p}$, then it would also be true identically in $\mathfrak{p}$, because the function (104) is a regular function in $\mathfrak{p}$ on $\mathcal{R}$ without any exceptions. Circling around $\mathfrak{p}$ on $\mathcal{R}$ increases $w(\mathfrak{p})$ by a period Ω; the argument in (104) increases by $\frac{p}{n}\Omega$. Letting $\Omega_1, \ldots, \Omega_p$ run through all systems of periods and letting n tend to infinity, then the values $\frac{p}{n}\Omega_1, \ldots, \frac{p}{n}\Omega_p$ converge to any system of complex numbers. However, $\theta(s)$ does not vanish identically as a function of the independent variables $s_1, \ldots, s_p$; therefore (104) cannot vanish identically in $\mathfrak{p}$ for n sufficiently large. This obviously holds independently from the choice of the constants $b_1, \ldots, b_p$. Henceforth, let now n be sufficiently large.

Avoiding finitely many exceptional points $\mathfrak{p}$, by (101) there are exactly n^{2p} different group of points Π determined to every group of points $\mathfrak{P}_0$; let $\pi_1, \ldots, \pi_p$ be one of them. Let $\chi(\mathfrak{p})$ be a non-constant rational function on the surface $\mathcal{R}$; let ρ be its order, $\nu^{(1)}, \ldots, \nu^{(\rho)}$ be its zeros, $\pi^{(1)}, \ldots, \pi^{(\rho)}$ be its poles. Then

$$\chi(\pi_1)\cdots\chi(\pi_p) = c\prod_{m=1}^{\rho}\frac{\theta(\sigma - w(\nu^{(m)}) + w(\mathfrak{a}))}{\theta(\sigma - w(\pi^{(m)}) + w(\mathfrak{a}))}, \tag{105}$$

where c is a constant. Letting $\mathfrak{p}$ run through a closed curve on $\mathcal{R}$ n-times, then by (103) the value of σ_k increases by a full period. Thus the ABEL functions of $\sigma_1, \ldots, \sigma_p$ as functions of the variable $\mathfrak{p}$ are certainly rational on that covering space $\mathcal{U}$ of the RIEMANN surface $\mathcal{R}$, on which all closed loops on $\mathcal{R}$, which are run through n-times, are again closed. To obtain $\mathcal{U}$ from $\mathcal{R}$, one has to perform the n^{2p} covering transformations on $\mathcal{R}$, which are obtained by running g_l-times through the return slices $\mathcal{A}_l$ and h_l-times through the return slices $\mathcal{B}_l$ ($l = 1, \ldots, p$; $g_l = 0, \ldots, n-1$; $h = 0, \ldots, n-1$). Next the order of the special function $\chi(\pi_1)\cdots\chi(\pi_p) = \Phi(\mathfrak{p})$ on $\mathcal{U}$ will be determined. For this purpose, one has to calculate the number of zeros of the function (104) on $\mathcal{U}$. More generally, consider $\theta\left(\frac{q}{n}w(\mathfrak{p}) + a\right)$ with q an integer and $a_1, \ldots, a_p$ arbitrary constants. The number of zeros of this function θ is equal to the change of $\frac{1}{2\pi i}\log\theta$ by running through the positively oriented boundary of the canonically sliced surface $\mathcal{U}$. Since $\mathcal{U}$ is composed of n^{2p} copies of $\mathcal{R}$, which are obtained by the covering transformations mentioned above, one only has to calculate the sum of the changes of $\frac{1}{2\pi i}\log\theta$ by running through the boundaries of the n^{2p} copies of the canonically sliced surface $\mathcal{R}$. Taking into consideration the equations

$$\theta(s + e_k) = \theta(s), \quad \theta(s + \tau_k) = e^{-\pi i \tau_{kk} - 2\pi i s_k}\theta(s),$$

hence

$$\theta\left(\frac{q}{n}s + qe_k\right) = \theta\left(\frac{q}{n}s\right), \quad \theta\left(\frac{q}{n}s + q\tau_k\right) = e^{-\pi i q^2\tau_{kk} - 2\pi i \frac{q^2}{n}s_k}\theta\left(\frac{q}{n}s\right),$$

it follows in the usual way that the slices $\mathcal{B}_l$ have no contribution, while n^{2p-2}-times the contribution $\frac{q^2}{n}\cdot n$ from the slices $\mathcal{A}_l$ is obtained, and this for $l = 1, \ldots, p$. The number of zeros is therefore exactly equal to $p\cdot q^2 n^{2p-2}$. As the right-hand side of the expression (105) for $\Phi(\mathfrak{p})$ contains in the denominator a factor θ exactly ρ-times, *the order of* $\Phi(\mathfrak{p})$ *on* $\mathcal{U}$ *is at most* $\rho p^3 n^{2p-2}$. For the purpose of this exposition it is essential that the exponent of n is not more than by 2 smaller than $2p$.

Put

$$\chi(\mathfrak{p}) = t - \frac{\alpha_1 x + \beta_1 y + \gamma_1}{\alpha_2 x + \beta_2 y + \gamma_2} \tag{106}$$

for arbitrary constants $t, \alpha_1, \beta_1, \gamma_1, \alpha_2, \beta_2, \gamma_2$, where $\alpha_1, \beta_1, \gamma_1$ are not proportional to $\alpha_2, \beta_2, \gamma_2$; here x and y denote two functions which generate the algebraic function field of $\mathcal{R}$, thus they are coordinates of the point $\mathfrak{p}$. Let h be the degree of the algebraic relation connecting x and y, then the order ρ of $\chi(\mathfrak{p})$ on $\mathcal{R}$ is equal to h. The order of $\chi(\pi_1) \cdots \chi(\pi_p) = \Phi(\mathfrak{p})$ on $\mathcal{U}$ is therefore at most hp^3n^{2p-2}. In (106) one chooses p different values $t_1, \dots, t_p$ for t; thereby $\Phi(\mathfrak{p})$ becomes

$$\Phi_k(\mathfrak{p}) = t_k^p + C_1 t_k^{p-1} + \cdots + C_p. \qquad (k = 1, \dots, p) \tag{107}$$

A second fact, which is subsequently of importance, is now:

Among $\Phi_1(\mathfrak{p}), \dots, \Phi_p(\mathfrak{p})$ no algebraic relation of degree smaller than $\frac{n^2}{hp^{2p+1}}$ exists identically in $\mathfrak{p}$.

Let $\nu_k^{(1)}, \dots, \nu_k^{(h)}$ be the t_k-points of $\frac{\alpha_1 x + \beta_1 y + \gamma_1}{\alpha_2 x + \beta_2 y + \gamma_2}$, then by (105) it holds

$$\Phi_k(\mathfrak{p}) = c \prod_{m=1}^{h} \frac{\theta(\sigma - w(\nu_k^{(m)}) + w(\mathfrak{a}))}{\theta(\sigma - w(\pi^{(m)}) + w(\mathfrak{a}))}. \qquad (k = 1, \dots, p) \tag{108}$$

Suppose now that there exists an algebraic relation of degree δ between the right-hand sides of (108) identically satisfied in $\mathfrak{p}$; thereby $\sigma_1, \dots, \sigma_p$ are the functions in the single variable $\mathfrak{p}$ determined by (103). This equation will be denoted for brevity by $G(\sigma) = 0$. It is not fulfilled if $\sigma_1, \dots, \sigma_p$ are independent variables; because in this case also $\pi_1, \dots, \pi_p$ would be independent variables, therefore also $C_1, \dots, C_p$ in (107), and finally also $\Phi_1, \dots, \Phi_p$ themselves. Setting

$$\sigma_k \equiv \frac{1}{n}(w_k(\mathfrak{p}_1) + \cdots + w_k(\mathfrak{p}_p)) + b_k, \qquad (k = 1, \dots, p)$$

then $G(\sigma)$ is identically 0 if the points $\mathfrak{p}_1, \dots, \mathfrak{p}_p$ all coincide, but is not identically 0 if the points are independent from each other. Now let q be the smallest natural number such that $G(\sigma)$ vanishes identically, whenever the points $\mathfrak{p}_1, \dots, \mathfrak{p}_q$ verify the condition $\mathfrak{p}_1 = \cdots = \mathfrak{p}_q$ and the points $\mathfrak{p}_{q+1}, \dots, \mathfrak{p}_p$ are varying freely. Let $1 < q \leq p$. Choose

$$\sigma_k \equiv \frac{1}{n}((q-1)w_k(\mathfrak{p}) + w_k(\mathfrak{p}_q) + \cdots + w_k(\mathfrak{p}_p)) + b_k. \tag{109}$$

For this argument $G(\sigma)$ does not vanish identically in $\mathfrak{p}, \mathfrak{p}_q, \dots, \mathfrak{p}_p$. Consider $G(\sigma)$ as a function of $\mathfrak{p}_q$; it is certainly rational on $\mathcal{U}$. The common denominator of the theta-quotients appearing in $G(\sigma)$ is

$$\left\{ \prod_{m=1}^{h} \theta(\sigma - w(\pi^{(m)}) + w(\mathfrak{a})) \right\}^{\delta}.$$

As we have already seen above, this expression vanishes on $\mathcal{U}$ exactly at δhpn^{2p-2} points $\mathfrak{p}_q$. On the other hand, $\theta(\sigma)$ is identically 0 if the variable $\mathfrak{p}_q$ in (109) is set to be equal to $\mathfrak{p}$. Circling on $\mathcal{R}$ changes $(q-1)w_k(\mathfrak{p}) + w_k(\mathfrak{p})$ into $(q-1)w_k(\mathfrak{p}) + w_k(\mathfrak{p}) + q\Omega_k$. Thus $G(\sigma)$ vanishes as a function of $\mathfrak{p}_q$ in $\mathfrak{p}$ and in those points which are obtained from $\mathfrak{p}$ by a covering transformation applied q-times. If d is the greatest common divisor of q and n, then $G(\sigma)$ has at least $(n:d)^{2p}$ zeros as a function of $\mathfrak{p}_q$ on $\mathcal{U}$. Therefore

$$\left(\frac{n}{d}\right)^{2p} \leq \delta hpn^{2p-2},$$

(110)
$$\delta \geq \frac{n^2}{d^{2p}hp}.$$

Since $d \leq q \leq p$, the statement follows. For the subsequent considerations it is once again essential that the exponent of n in (110) is not smaller than 2.

§5. Equations of arbitrary positive genus.

Let $f(x,y) = 0$ be of genus $p \geq 1$. Consider all systems of p points $x_1, y_1; x_2, y_2; \ldots; x_p, y_p$, with the property that all their rational symmetric combinations with coefficients in $\mathfrak{K}$ again belong to $\mathfrak{K}$. The systems $s_1, \ldots, s_p$ associated to these groups of points according to (99) determine, by WEIL's theorem, a finite module $\mathfrak{M}$. If the systems $s_1^{(1)}, \ldots, s_p^{(1)}; \ldots; s_1^{(q)}, \ldots, s_p^{(q)}$ are a basis of the module, then in analogy with (95)

(111)
$$s_k \equiv n\sigma_k + c_k, \qquad (k = 1, \ldots, p)$$

where $\sigma_1, \ldots, \sigma_p$ are elements of the module and $c_1, \ldots, c_p$ are one of finitely many elements of $\mathfrak{M}$; it suffices to restrict c_k to the values $n_1 s_k^{(1)} + \cdots + n_q s_k^{(q)}$, where $n_1, \ldots, n_q$ belong to the numbers $0, \ldots, n-1$.

Assume now that $f(x,y) = 0$ has infinitely many solutions x, y in $\mathfrak{K}$ with x an integer. The point x, y shall be denoted by $\mathfrak{p}$. The values of $s_1, \ldots, s_p$ corresponding to the points $\mathfrak{p}, \ldots, \mathfrak{p}$ are in $\mathfrak{M}$. From the set of the points $\mathfrak{p}$ an infinite subset is taken for which in (111) a fixed system $c_1, \ldots, c_p$ appears. The following considerations are made for those $\mathfrak{p}$. Excluding finitely many of the $\mathfrak{p}$, the n^{2p} groups of points $\pi_1, \ldots, \pi_p$ are uniquely determined. Let ξ_l, η_l be the coordinates of the point π_l $(l = 1, \ldots, p)$. By (111), among the n^{2p} groups of points $\pi_1, \ldots, \pi_p$ there is one group for which all rational symmetric functions of the p pairs ξ_l, η_l with coefficients in $\mathfrak{K}$ lead to a value still belonging to $\mathfrak{K}$. If one chooses for the constants $\alpha_1, \ldots, \gamma_2, t_1, \ldots, t_p$ in the previous paragraph values of $\mathfrak{K}$, then in particular the quantities $\Phi_1(\mathfrak{p}), \ldots, \Phi_p(\mathfrak{p})$ all belong to $\mathfrak{K}$.

The field of all rational symmetric functions of the p pairs ξ_l, η_l is contained in the algebraic function field belonging to $\mathcal{U}$; its RIEMANN surface is therefore either $\mathcal{U}$ itself or a surface $\mathcal{U}'$ to which $\mathcal{U}$ is a covering. Let $\mathcal{U}'$ consist of ν copies of $\mathcal{R}$; then ν is a divisor of n^{2p}. It shall be shown that for a suitable choice of the constants $\alpha_1, \dots, \gamma_2, t_1, \dots, t_p$ the two functions $\Phi_1(\mathfrak{p})$ and $\Phi_k(\mathfrak{p})$ for $k = 2, \dots, p$ generate the field belonging to $\mathcal{U}'$. To each point $\mathfrak{p}$ there is exactly one group of points $\pi_1, \dots, \pi_p$ of $\mathcal{U}'$ corresponding to $\mathfrak{p}$; different systems $\pi_1, \dots, \pi_p$ are associated to different $\mathfrak{p}$. It follows

$$\text{(112)} \qquad \begin{aligned} \Phi_k(\mathfrak{p}) &= \prod_{l=1}^{p} \left(t_k - \frac{\alpha_1 \xi_l + \beta_1 \eta_l + \gamma_1}{\alpha_2 \xi_l + \beta_2 \eta_l + \gamma_2} \right) \\ &= t_k^p + C_1 t_k^{p-1} + \cdots + C_p. \qquad (k = 1, \dots, p) \end{aligned}$$

Let a denote a value, which is not assumed by Φ_1 more than once. If the order of Φ_1 on $\mathcal{U}'$ is λ, then λ different points $\mathfrak{p}$ satisfy the equation $\Phi(\mathfrak{p}) = a$. For these λ points, one considers the system of the coefficients $C_1, \dots, C_p$ on the right-hand side of (112); it may happen that for two different points among the λ points $\mathfrak{p}$ the systems $C_1, \dots, C_p$ coincide. In this case however, also the system of the p expressions $(\alpha_1 \xi_l + \beta_2 \eta_l + \gamma_1) : (\alpha_2 \xi_l + \beta_2 \eta_l + \gamma_2)$ $(l = 1, \dots, p)$ for these two $\mathfrak{p}$ must coincide up to the order. But this can not identically hold in $\alpha_1, \dots, \gamma_2$, since then the two corresponding systems of points ξ_l, η_l $(l = 1, \dots, p)$ would also coincide. Hence, one can choose $\alpha_1, \dots, \gamma_2$ so that for the λ solutions of $\Phi_1 = a$ the λ systems $C_1, \dots, C_p$ are different; in turn choose t_k for $k = 2, \dots, p$ so that the values of $t_k^p + C_1 t_k^{p-1} + \cdots + C_p$ for these λ systems $C_1, \dots, C_p$ are also all different from each other. Then Φ_1 and Φ_k generate the field belonging to $\mathcal{U}'$. According to the order of Φ_k and, more generally, of any linear combination on $\mathcal{U}$ of $\Phi_1, \dots, \Phi_p$, with constant coefficients, is at most equal to $hp^3 n^{2p-2}$; hence on $\mathcal{U}'$ the order is at most equal to $(hp^3 n^{2p-2}) : (n^{2p}/\nu) = \frac{hp^3 \nu}{n^2}$. The degree of the algebraic equation satisfied by Φ_1 and Φ_k is at most $hp^3 \nu : n^2$. The coefficients of this equation belong to $\mathfrak{K}$.

Let ζ_0 be the common denominator of the p numbers $\Phi_1, \dots, \Phi_p$, *i.e.* $\Phi_1 = \zeta_1 : \zeta_0, \dots, \Phi_p = \zeta_p : \zeta_0$ with coprime $\zeta_0, \zeta_1, \dots, \zeta_p$. Moreover, let μ_k be the common denominator of Φ_1 and Φ_k for $k = 2, \dots, p$. One forms the expression

$$\text{(113)} \qquad A = \prod_{k=2}^{p} N(|\mu_k \Phi_1| + |\mu_k \Phi_k| + |\mu_k|),$$

where the norm N is taker over the l conjugate fields of $\mathfrak{K}$. Obviously, the product $\mu_2 \cdots \mu_p$ is a multiple of ζ_0 and consequently

$$\text{(114)} \qquad \begin{aligned} A &\geq N\left(|\zeta_0| \prod_{k=2}^{p} (|\Phi_1| + |\Phi_k| + 1) \right) \\ &\geq N(|\zeta_0 \Phi_1| + |\zeta_0 \Phi_2| + \cdots + |\zeta_0 \Phi_p| + |\zeta_0|) \\ &= N(|\zeta_0| + \cdots + |\zeta_p|). \end{aligned}$$

The result of §2 is now applied to the function x, which is rational on $\mathcal{R}$, hence also on $\mathcal{U}'$, and therefore belongs to the function field generated by Φ_1 and Φ_k. By assumption x is an integer. Let the order of x on $\mathcal{R}$ be $\rho \leq h$; since $\mathcal{U}'$ consists of ν copies of $\mathcal{R}$, the order of x on $\mathcal{U}'$ exactly equals $\rho\nu$. Moreover, x does not assume any value on $\mathcal{U}'$ more than ρ-times. The number g of §2 has here the value $\rho\nu$, moreover, for the numbers h and r one has the upper bounds $hp^3\nu : n^2$ and ρ here. By §2, a subsequence of the $\mathfrak{p}$ converges to a pole $\mathfrak{p}_0$ of x on $\mathcal{U}'$, and in these $\mathfrak{p}$ the estimate (91) holds for every rational function $\phi(\mathfrak{p})$ on $\mathcal{U}'$ that vanishes at $\mathfrak{p}_0$, namely

$$|\phi(\mathfrak{p})| < c_9 N(|\mu_k \Phi_1| + |\mu_k \Phi_2| + |\mu_k|)^{-\frac{n^2}{hlp^3}+\epsilon}$$

for $k = 2, \ldots, p$. Multiplying these inequalities, it follows for $p > 1$ by using (113) and (114)

$$|\phi(\mathfrak{p})| < c_{11} N(|\zeta_0| + \cdots + |\zeta_p|)^{-\kappa n^2+\epsilon}, \tag{115}$$

where κ denotes the number $1 : hlp^3(p-1)$, which does not depend on n, ϵ is any positive number and c_{11} does not depend on $\mathfrak{p}$.

It can be assumed that the values $\Phi_1(\mathfrak{p}_0), \ldots, \Phi_p(\mathfrak{p}_0)$ are all finite, because otherwise it would just be sufficient to modify the parameters $\alpha_1, \ldots, \gamma_2$ in (112). Fixing $\Phi_k(\mathfrak{p}_0) = \omega_k$, then by (115) in particular

$$|\Phi_k(\mathfrak{p}) - \Phi_k(\mathfrak{p}_0)| = \left|\frac{\zeta_k}{\zeta_0} - \omega_k\right| < c_{11} N(|\zeta_0| + \cdots + |\zeta_p|)^{-\kappa n^2+\epsilon} \tag{116}$$

for $k = 1, \ldots, p$.[18] At this point one cannot just proceed with the same argument as in §3; because the degree of the algebraic number ω_k could be of order n^{2p}, since ω_k was determined through the n-slicing of the periods of the ABEL functions; and n^{2p} grows for $p \geq 2$ not less than the square of the exponent of the right-hand side of (116). However, (116) contains p approximation assertions and the approximated numbers ω_k, as will be shown in §4, are "sufficiently" independent from each other. This aspect makes it possible to successfully use the method from the introduction and to deduce a contradiction to (116).

Subsequently, the following lemma will be needed:

Let δ be smaller than the two numbers κn^2 and $\frac{n^2}{hp^{2p+1}}$. Then no algebraic relation of degree δ with coefficients in $\mathfrak{K}$ involving the numbers $\omega_1, \ldots, \omega_p$

[18] FOOTNOTE BY THE EDITORS: It is here that the role of Roth's theorem (unavailable at the time when Siegel wrote the paper) is replaced by simultaneous approximations expressed by these p inequalities. The needed result shall be proved below.

holds.[19] In other words: No polynomial of degree δ in $\Phi_1(\mathfrak{p}), \ldots, \Phi_p(\mathfrak{p})$ with coefficients in $\mathfrak{K}$ has $\mathfrak{p} = \mathfrak{p}_0$ as root.

Let $G(\mathfrak{p})$ be a polynomial of degree δ in $\Phi_1(\mathfrak{p}), \ldots, \Phi_p(\mathfrak{p})$ with integral coefficients from $\mathfrak{K}$, which vanishes at $\mathfrak{p} = \mathfrak{p}_0$. According to §4, it can not vanish identically in $\mathfrak{p}$, because δ is smaller than $\frac{n^2}{hp^{2p+1}}$. Using (115), for infinitely many $\mathfrak{p} \to \mathfrak{p}_0$ one has

$$|G(\mathfrak{p})| < c_{11} N(|\zeta_0| + \cdots + |\zeta_p|)^{-\kappa n^2+\epsilon}. \tag{117}$$

Now $\zeta_0^\delta G(\mathfrak{p})$ has an upper bound of the form $c_{12}(|\zeta_0| + \cdots + |\zeta_p|)^\delta$; on the other hand $\zeta_0^\delta G(\mathfrak{p})$ is integral and $\neq 0$ provided that $\mathfrak{p}$ lies sufficiently near to $\mathfrak{p}_0$; the norm of this number is therefore in absolute value at least 1. This gives by (117), the inequality

$$1 < c_{11} N(|\zeta_0| + \cdots + |\zeta_p|)^{-\kappa n^2+\epsilon} |\zeta_0|^\delta c_{13} \frac{N(|\zeta_0| + \cdots + |\zeta_p|)^\delta}{(|\zeta_0| + \cdots + |\zeta_p|)^\delta},$$

and therefore also

$$N(|\zeta_0| + \cdots + |\zeta_p|)^{\kappa n^2-\epsilon-\delta} < c_{14},$$

which is a contradiction to $\delta < \kappa n^2$. This proves the lemma.

§6. An application of the approximation method.

Let the field $\mathfrak{K}'$ obtained by adjoining the numbers $\omega_1, \ldots, \omega_p$ to $\mathfrak{K}$ be of degree d. In order to determine the numbers $\omega_1, \ldots, \omega_p$ one just has to solve a single algebraic equation of degree n^{2p} with coefficients in $\mathfrak{K}$, indeed the equation on which the n-slicing of the periods of the ABEL functions depends. Therefore $d \leq ln^{2p}$. Without loss of generality it can be supposed that ω_1 generates the field $\mathfrak{K}'$. If this were not the case, it would be sufficient to replace $\Phi_1(\mathfrak{p})$ by a suitable fixed linear combination of $\Phi_1(\mathfrak{p}), \ldots, \Phi_p(\mathfrak{p})$ with coefficients in $\mathfrak{K}$, which would leave the inequalities (116) unmodified, apart from the value of the constant c_{11}. Moreover, we may assume that $\omega_1, \ldots, \omega_p$ are integers, because otherwise one would only have to multiply $\Phi_1, \ldots, \Phi_p$ by the common denominator of these numbers. The number ω_1 shall be abbreviated in the sequel by ω.

[19] FOOTNOTE BY THE EDITORS: The approximation theorem that Siegel derives is adjusted just for his present needs. To our knowledge, he did not publish a general statement in this direction. Actually, Siegel is referring to the paper translated here only at very rare occasions, namely in [C.L. Siegel, Die Gleichung $ax^n - by^n = c$, Math. Ann. **114** (1937), 57-68] and in [C.L. Siegel, Einige Erläuterungen zu Thues Untersuchungen über Näherungswerte algebraischer Zahlen und diophantische Gleichungen, Nachr. Akad. Wiss. Göttingen Math.-Phys. Kl. II 1970, 169–195]. We also mention [C.L. Siegel, Zur Theorie der quadratischen Formen, Nachr. Akad. Wiss. Göttingen Math.-Phys. Kl. II 1972, 21–46], in which he is mentioning work of Thue, Baker, Davenport, Roth and Schidlowski.

Let λ be the smaller of the two numbers $\kappa = 1 : hlp^3(p-1)$ and $1 : hp^{2p+1}$. Let

$$n > \lambda^{-\frac{1}{2}}$$

and

(118) $$1 \leq \delta < \lambda n^2.$$

The number of all the monomials in $\omega_1, \ldots, \omega_p$ of degree $\leq \delta$ is exactly given by

(119) $$\binom{\delta + p}{p} = m + 1;$$

they shall be denoted by $\alpha_0, \ldots, \alpha_m$ in some order. Among these, the number 1 appears. Because of (118), it follows from the lemma of the previous paragraph that the numbers $\alpha_0, \ldots, \alpha_m$ are linearly independent in the field $\mathfrak{K}$. Now, let $P_0(x), \ldots, P_m(x)$ be polynomials in x of degree q with rational integer coefficients. In order to make the polynomial

$$P(x) = \alpha_0 P_0(x) + \cdots + \alpha_m P_m(x)$$

vanish at least of order b at $x = \omega$, the equations

(120) $$\alpha_0 P_0^{(k)}(\omega) + \cdots + \alpha_m P_m^{(k)}(\omega) = 0,$$

whose left-hand sides are obtained by differentiating k times $P(x)$, must be valid for $k = 0, 1, \ldots, b-1$. These are b linear homogeneous equations for the $(m+1)(q+1)$ coefficients of $P_0(x), \ldots, P_m(x)$. If one expresses all the numbers of $\mathfrak{K}'$ that appear in these equations in terms of a basis with respect to the field of rational numbers, then every equation splits into d new equations with rational integer coefficients; hence one has bd equations in $(m+1)(q+1)$ unknowns.

For every natural a the number $\binom{a}{k}$ is an integer. Therefore the rational integer coefficients of the d equations obtained from (120) have the common divisor $k!$; divide them through by this term. Taking into account that $\binom{a}{k} < 2^a$, it follows that all rational integer coefficients of the bd equations are smaller than c_{15}^q in absolute value. Here c_{15} is a number, which does not depend on b and q; the same is assumed to hold for the numbers $c_{16}, \ldots, c_{26}$ below.

Let now

(121) $$b > 2(m+1)^2(2m+1)d.$$

One fixes q through

(122) $$q = \left[\frac{b}{m+1}\left(d + \frac{1}{2m+1}\right)\right] - 1.$$

Moreover, one puts

$$\theta = \frac{(m+1)(q+1)}{bd} - 1,$$

then using (121) and (122)

$$\frac{1}{(2m+2)d} < \theta \le \frac{1}{(2m+1)d}. \tag{123}$$

Now the second lemma from the introduction is applied; the numbers that are denoted by n and m there have here the values $(m+1)(q+1)$ and bd, moreover A has c_{15}^q as upper bound. Therefore it is possible to determine the coefficients of $P_0(x), \ldots, P_m(x)$ as rational integer numbers, that are not all 0, that are smaller than $1 + \{(m+1)(q+1)c_{15}^q\}^{\frac{1}{\theta}}$ in absolute value, and that satisfy the equations (120) for $k = 0, 1, \ldots, b-1$. By taking (122) and (123) into account, the bound satisfies

$$1 + \{(m+1)(q+1)c_{15}^q\}^{\frac{1}{\theta}} < c_{16}^b. \tag{124}$$

Among the polynomials $P_0(x), \ldots, P_m(x)$ some can be linearly dependent from each other, then they are also linearly dependent over to the field of rational numbers. Let $\mu + 1$ be the number of linearly independent ones among those; one may assume that $P_0(x), \ldots, P_\mu(x)$ are linearly independent. Then one has

$$P(x) = \beta_0 P_0(x) + \cdots + \beta_\mu P_\mu(x),$$

where $\beta_0, \ldots, \beta_\mu$ denote linear homogeneous combinations of $\alpha_0, \ldots, \alpha_m$ with rational coefficients. Since $\alpha_0, \ldots, \alpha_m$ are linearly independent in $\mathfrak{K}$, $\beta_0, \ldots, \beta_\mu$ are all non-zero numbers. Let

$$W(x) = \left| P_l^{(k)}(x) \right|,$$

where the row-index k and the column-index l assume the values $0, \ldots, \mu$, be the WRONSKIan determinant of the polynomials $P_0, \ldots, P_\mu$; it does not identically vanish in x and is a polynomial of degree $(\mu+1)q$ with rational coefficients. Let $W_{kl}(x)$ be the cofactor of $P_l^{(k)}$ in W. From the $\mu + 1$ equations

$$P^{(k)}(x) = \beta_0 P_0^{(k)}(x) + \cdots + \beta_\mu P_\mu^{(k)}(x) \qquad (k = 0, \ldots, \mu)$$

it follows that

$$\beta_k W(x) = \sum_{l=0}^{\mu} W_{kl}(x) P^{(l)}(x), \qquad (k = 0, \ldots, \mu) \tag{125}$$

and this holds identically in $\beta_0, \ldots, \beta_\mu$. Now, $P(x)$ vanishes at $x = \omega$ at least with order b, consequently $P^{(l)}(x)$ at least with order $b - l$. Since $\beta_k \neq 0$, then the polynomial $W(x)$ vanishes at $x = \omega$, according to (125), at least with order $b - \mu$. This is also true for all the d conjugates of ω. Therefore

$$(\mu + 1)q - d(b - \mu) = g \geq 0, \tag{126}$$

and $W(x)$ vanishes for any arbitrary value ξ, different from the conjugates of ω, at most with order g. But from (121), (122) and (126) it follows that

$$\begin{aligned} 0 \leq g &< \frac{\mu + 1}{m + 1} bd + \frac{b}{2m + 1} + md - db \\ &< b\left(\frac{\mu - m}{m + 1} d + \frac{1}{2m + 1} + \frac{m}{2(m + 1)^2(2m + 1)}\right) \\ &< \frac{b}{m + 1}\{(\mu - m)d + 1\}, \end{aligned}$$

then $\mu = m$ and

$$g < \frac{b}{m + 1}. \tag{127}$$

Since $\mu = m$ all the polynomials $P_0(x), \ldots, P_m(x)$ are linearly independent from each other and one can consequently replace α_k by β_k $(k = 0, \ldots, m)$.

Let ξ be different from the d conjugates of ω. Assume that the WRONSKIan determinant $W(x)$ vanishes at $x = \xi$ of order $\gamma \leq g$. Then $W^{(\gamma)}(\xi) \neq 0$. On the other hand by (125) the number $\alpha_k W^{(\gamma)}(\xi)$ is for $k = 0, \ldots, m$ a homogeneous linear combination of $P(\xi), P'(\xi), \ldots, P^{(m+\gamma)}(\xi)$, indeed this holds identically in $\alpha_0, \ldots, \alpha_m$. Therefore among the $m + \gamma + 1$ expressions $P(\xi), P'(\xi), \ldots, P^{(m+\gamma)}(\xi)$, $m + 1$ are linearly independent from each other if considered as homogeneous linear functions of $\alpha_0, \ldots, \alpha_m$. Let those be the expressions $P^{(k)}(\xi)$ for $k = k_0, \ldots, k_m$. Dividing through by $k!$ one obtains the values

$$\begin{aligned} Q_0(\xi) &= P_{00}(\xi)\alpha_0 + \cdots + P_{0m}(\xi)\alpha_m \\ &\vdots \\ Q_m(\xi) &= P_{m0}(\xi)\alpha_0 + \cdots + P_{mm}(\xi)\alpha_m. \end{aligned} \tag{128}$$

The determinant $|P_{kl}(\xi)|$ is not 0. The $P_{kl}(\xi)$ are polynomials of degree q in ξ; their coefficients are rational integers and, by (124), smaller than c_{17}^b in absolute value. Moreover, $Q_0(x), \ldots, Q_m(x)$ vanish at $x = \omega$ at least with order $b' = b - m - \gamma$. Using (121) and (127), one gets

$$b' \geq b - m - g > b\left(1 - \frac{m}{2(m + 1)^2(2m + 1)} - \frac{1}{m + 1}\right) > \frac{b}{4}. \tag{129}$$

From the formula

$$Q_k(\xi) = \frac{1}{2\pi i}(\xi - \omega)^{b'} \int \frac{Q_k(t)}{(t - \omega)^{b'}} \frac{dt}{t - \xi},$$

where the integral is e.g. extended over the circle $|t - \xi| = 1$, one deduces the estimate

$$(130) \qquad |Q_k(\xi)| < c_{18}^b \, |\xi - \omega|^{b'} \, (1 + |\xi|)^{q-b'}. \qquad (k = 0, \dots, m)$$

Among the infinitely many points $\mathfrak{p}$ with limit $\mathfrak{p}_0$, for which (116) holds, one now takes two, say $\mathfrak{p}$ and $\mathfrak{p}'$. Let the values $\Phi_k(\mathfrak{p}')$ be denoted by $\zeta_k' : \zeta_0'$, for $k = 1, \dots, p$ with $\zeta_0', \dots, \zeta_p'$ coprime; moreover put

$$(131) \qquad \log N(|\zeta_0| + \cdots + |\zeta_p|) = \Lambda, \quad \log N(|\zeta_0'| + \cdots + |\zeta_p'|) = \Lambda'.$$

If $\mathfrak{p}'$ lies sufficiently near to $\mathfrak{p}_0$, then $\zeta_1' : \zeta_0'$ is different from ω and its conjugates; in fact, if $\zeta_1' : \zeta_0' = \omega$, then ω would be a number from $\mathfrak{K}$, and using (118) one would find a contradiction to §5; and, for $\mathfrak{p}'$ sufficiently near to $\mathfrak{p}_0$, $\Phi_1(\mathfrak{p}')$ can also not be equal to one of the conjugates of $\omega = \Phi_1(\mathfrak{p}_0)$ different from ω. Consequently, it is possible to choose $\xi = \zeta_1' : \zeta_0'$. According to (116), (130) and (131) it holds

$$\left| Q_k\left(\frac{\zeta_1'}{\zeta_0'}\right) \right| < c_{19}^b e^{-\Lambda'(\kappa n^2 - \epsilon) b'}, \qquad (k = 0, \dots, m)$$

and then by (129)

$$(132) \qquad \left| Q_k\left(\frac{\zeta_1'}{\zeta_0'}\right) \right| < c_{19}^b e^{-\kappa_1 b \Lambda' n^2},$$

where κ_1 denotes a positive constant, which does not depend on b and n.

In the right-hand side of (128) one replaces the monomials $\alpha_0, \dots, \alpha_m$ in $\omega_1, \dots, \omega_p$ by the monomials $A_0, \dots, A_m$ obtained analogously in $\zeta_1 : \zeta_0, \dots, \zeta_p : \zeta_0$. Thereby, instead of $Q_0(\xi), \dots, Q_m(\xi)$ the numbers

$$\begin{aligned} Z_0 &= P_{00}(\xi) A_0 + \cdots + P_{0m}(\xi) A_m \\ &\vdots \\ Z_m &= P_{m0}(\xi) A_0 + \cdots + P_{mm}(\xi) A_m \end{aligned}$$

appear, with $\xi = \zeta_1' : \zeta_0'$. These numbers all belong to $\mathfrak{K}$. The determinant $|P_{kl}(\xi)|$ is $\neq 0$ and among the numbers $A_0, \dots, A_m$ the number 1 appears. Therefore at least one of the numbers $Z_0, \dots, Z_m$ is different from 0, say the number $Z_0 = Z$. Now, every $P_{kl}(\xi)$ is a polynomial of degree q in ξ, moreover the monomials $A_0, \dots, A_m$ have at most degree δ. Consequently, the number

$$(\zeta_0')^q \zeta_0^\delta Z = \Gamma$$

is integral. The conjugates of Γ have as an upper bound the conjugates of

$$c_{20}^b (|\zeta_0'| + \cdots + |\zeta_p'|)^q (|\zeta_0| + \cdots + |\zeta_p|)^\delta$$

and Γ itself is bounded from above by

$$|Z|\,(|\zeta_0'| + \cdots + |\zeta_p'|)^q (|\zeta_0| + \cdots + |\zeta_p|)^\delta.$$

Therefore using (131), it holds

$$1 \le |N\Gamma| < c_{21}^b e^{\Lambda' q + \Lambda\delta} |Z|. \tag{133}$$

One now uses the estimate

$$\left|\left(\frac{\zeta_1}{\zeta_0}\right)^{l_1} \cdots \left(\frac{\zeta_p}{\zeta_0}\right)^{l_p} - \omega_1^{l_1} \cdots \omega_p^{l_p}\right| < c_{22}\left(\left|\frac{\zeta_1}{\zeta_0} - \omega_1\right| + \cdots + \left|\frac{\zeta_p}{\zeta_0} - \omega_p\right|\right)$$

and once more (116); it follows

$$\left|Z_k - Q_k\left(\frac{\zeta_1'}{\zeta_0'}\right)\right| < c_{23}^b e^{-\Lambda(\kappa n^2 - \epsilon)} \qquad (k = 0, \ldots, m)$$

and in combination with (132)

$$|Z| < c_{24}^b (e^{-\kappa_1 n^2 \Lambda} + e^{-\kappa_1 n^2 b \Lambda'}). \tag{134}$$

The still arbitrary number b is now fixed as

$$b = \left[\frac{\Lambda}{\Lambda'}\right];$$

since b is assumed to satisfy the inequality (121), this implies a limitation for the choice of $\mathfrak{p}$, which can however be satisfied, since Λ is unbounded as $\mathfrak{p} \to \mathfrak{p}_0$. Then $b\Lambda' \le \Lambda < (b+1)\Lambda'$, and using (133) and (134) one gets

$$\Lambda'(q + (b+1)\delta - \kappa_1 n^2 b) + c_{25} b > 0. \tag{135}$$

For δ one chooses the value

$$\delta = \left[n^{\frac{2p}{p+1}}\right]; \tag{136}$$

in order to satisfy (118), n must be so big that

$$\left[n^{\frac{2p}{p+1}}\right] < \lambda n^2. \tag{137}$$

Because of $d \le l n^{2p}$, it follows by (119), (122) and (136) that

$$q < \kappa_2 b n^{2p} : \left(n^{\frac{2p}{p+1}}\right)^p = \kappa_2 n^{\frac{2p}{p+1}} b,$$

where κ_2 does not depend on n and b, moreover,

$$(b+1)\delta < \kappa_3 n^{\frac{2p}{p+1}} b$$

with an analogous meaning of κ_3. The factor of Λ' in (135) is smaller than

$$b\{(\kappa_2+\kappa_3)n^{\frac{2p}{p+1}} - \kappa_1 n^2\} = -c_{26} b.$$

Now one chooses n so big that (137) is satisfied and that the number

$$c_{26} = n^2 \left\{\kappa_1 - (\kappa_2+\kappa_3)n^{\frac{2}{p+1}}\right\}$$

is positive, and subsequently $\mathfrak{p}'$ so that $\Lambda' > c_{25} : c_{26}$. This results in a contradiction to (135).

This proves the theorem also in the case $p > 1$.

§7. Cubic forms with positive discriminant.

The investigations of the previous paragraphs open the possibility to determine a bound for the number of the solutions to the equation $f(x,y) = 0$ as a function of the coefficients of f explicitly, provided that the equation has just a finite number of solutions. One can conjecture that it is even possible to find a bound which only depends on the number of coefficients; however, this seems rather hard to prove.[20] Support to this conjecture gives the very special method that will be developed below.

By the studies of B. DELAUNAY and NAGELL it has been established that a cubic form with rational integer coefficients and negative discriminant assumes at most five times the value 1 for rational integer values of the variables x and y. The proofs of the two authors make use of the theory of units and it seems that they cannot be generalized to the cases of cubic forms with positive discriminant and with values different from 1 or 3. But this becomes possible through the approach, which THUE made at the beginning of his studies on diophantine equations, namely to use the continued fraction for the function $(1-x)^\alpha$; one just has to slightly refine THUE's estimates.[21]

[20] FOOTNOTE BY THE EDITORS: Siegel raises the question whether there are bounds for the number of solutions depending only on the number of nonzero coefficients. This is not true in such a general form, but several authors as Bombieri, Schmidt, Evertse and Schlickewei proved certain uniform results in Siegel's sought direction; see the article on integral points, which we have included after the translation, for references.

[21] FOOTNOTE BY THE EDITORS: Siegel later made a historical correction in [C.L. Siegel, Einige Erläuterungen zu Thues Untersuchungen üuber Annäherungswerte algebraischer Zahlen und diophantische Gleichungen, Nachr. Akad. Wiss. Göttingen Math.-Phys. Kl. II 1970, 169–195]; see also "Berichtigungen und Bemerkungen" on p. 339 in [C.L. Siegel, Gesammelte Abhandlungen. Band IV. Edited by K. Chandrasekharan and H. Maass. With corrections to the first three volumes. Springer-Verlag, Berlin-New York, 1979]. There he says that he was mistaken in assuming that Thue came to this when working with the continued fraction for $(1-z)^v$ and that Thue was just working on this several years later without obtaining prominent results.

Let m and n be non-negative rational integer numbers and let F be the hypergeometric function.[22] Then it is

$$\begin{aligned}(1-z)^{\alpha}F(-n+\alpha,-m,-m-n,z)-F(-m-\alpha,-n,-m-n,z)\\=\lambda z^{m+n+1}F(n-\alpha+1,m+1,m+n+2,z)\end{aligned}$$

with

$$\lambda=(-1)^{n-1}\binom{m+\alpha}{m+n+1}:\binom{m+n}{n}.$$

This is considered now in the particular case when $m=n-g$ and $g=0,1$. One defines

$$\begin{aligned}A_g&=F(-n+\alpha,-n+g,-2n+g,z)\\B_g&=F(-n-\alpha+g,-n,-2n+g,z)\\R_g&=\lambda_g z^{2n-g+1}F(n-\alpha+1,n-g+1,2n-g+2,z)\\\lambda_g&=(-1)^{n-g}\binom{n+\alpha-g}{2n-g+1}:\binom{2n-g}{n},\end{aligned}$$

then one gets

$$(1-z)^{\alpha}A_g(z)-B_g(z)=R_g(z);\qquad (g=0,1)$$

and since $A_g(z)$ is a polynomial of degree $n-g$ and $B_g(z)$ is a polynomial of degree n, the expression equals

$$A_0B_1-A_1B_0=-A_0R_1+A_1R_0=-\lambda_1 z^{2n}+\cdots=-\lambda_1 z^{2n}, \tag{138}$$

hence is $\neq 0$ for $z\neq 0$ and $\alpha\neq -n+1,-n+2,\ldots,n$.

By RIEMANN for $n\to\infty$

$$\begin{aligned}B_g&\sim(1-z)^{\frac{\alpha}{2}-\frac{1}{4}}\left(\frac{\sqrt{1-z}+1}{2}\right)^{2n+1-g}\\R_g&\sim-2\sin\pi\alpha\,(1-z)^{\frac{\alpha}{2}-\frac{1}{4}}\left(\frac{\sqrt{1-z}-1}{2}\right)^{2n+1-g},\end{aligned}$$

[22] FOOTNOTE BY THE EDITORS: Hypergeometric function were used first by Thue, then also by Baker, Beukers and others, to obtain explicit results in Diophantine approximation to roots and other algebraic numbers.

and this holds true uniformly in every closed region of the z-plane, which does not contain the points 1 and ∞. Then also for arbitrary z' it holds

$$(1-z)^{\alpha}A_g - (1-z')^{\alpha}B_g = R_g + (1-(1-z')^{\alpha})B_g$$

(139)
$$\sim (1-z)^{\frac{\alpha}{2}-\frac{1}{4}}\left\{-2\sin\pi\alpha\left(\frac{\sqrt{1-z}-1}{2}\right)^{2n+1-g}\right.$$
$$\left.+(1-(1-z')^{\alpha})\left(\frac{\sqrt{1-z}+1}{2}\right)^{2n+1-g}\right\}.$$

Herein by (138), the left-hand side is not 0 when either $g=0$ or $g=1$, provided that z is different from 0 and 1 and that α is not a rational integer number. In the sequel, let $\alpha=\frac{1}{3}$.

By Theorem 5 in the first part of this exposition, the common denominator h of the coefficients of A_0, A_1, B_0, B_1 is at most equal to γ_1^n, where γ_1 denotes a positive constant.

Let

$$\phi(x,y) = a_0x^3 + a_1x^2y + a_2xy^2 + a_3y^3$$

be a cubic form with rational integer coefficients a_0, a_1, a_2, a_3 and with positive discriminant

$$d = a_1^2a_2^2 - 4a_0a_2^3 - 4a_1^3a_3 - 27a_0^2a_3^2 + 18a_0a_1a_2a_3.$$

Between the covariants

$$\psi(x,y) = -\frac{1}{4}\left(\frac{\partial^2\phi}{\partial x^2}\frac{\partial^2\phi}{\partial y^2} - \frac{\partial^2\phi}{\partial x\partial y}\frac{\partial^2\phi}{\partial x\partial y}\right)$$
$$\chi(x,y) = \frac{\partial\phi}{\partial x}\frac{\partial\psi}{\partial y} - \frac{\partial\phi}{\partial y}\frac{\partial\psi}{\partial x}$$

and ϕ the following identity holds

(140)
$$4\psi^3 = \chi^2 + 27d\phi^2.$$

The form $\psi(x,y)$ is quadratic and has negative discriminant $-3d$. From (140) it follows in a well known way that

(141)
$$\psi = \xi\eta, \quad \chi = \xi^3 + \eta^3, \quad \phi = \frac{\xi^3-\eta^3}{3\sqrt{-3d}},$$

(142)
$$\xi^3 = \frac{1}{2}(\chi + 3\phi\sqrt{-3d}), \quad \eta^3 = \frac{1}{2}(\chi - 3\phi\sqrt{-3d}),$$

where ξ and η are linear functions in x and y. There are two constants λ and μ so that $\xi : \lambda$ and $\eta : \mu$ are linear functions in x and y with coefficients belonging to the imaginary quadratic field $\mathfrak{K}$ generated by $\sqrt{-3d}$. If x and y are rational integers, then so are ϕ, ψ, χ; then (141) gives that ξ^3 and η^3 are conjugate integer numbers from $\mathfrak{K}$; and the numbers $\xi : \lambda$ and $\eta : \mu$ also lie in $\mathfrak{K}$.

Now let the diophantine equation

$$\phi(x, y) = k \tag{143}$$

be given, where k is a fixed natural number. For every solution x, y of this equation one forms the numbers ξ^3 and η^3 according to (142) and sets

$$\frac{3k\sqrt{-3d}}{\xi^3} = z.$$

Then by (141) and (143) one gets

$$\left(\frac{\eta}{\xi}\right)^3 = 1 - z,$$

hence

$$\frac{\eta}{\xi} = \epsilon(1 - z)^{\frac{1}{3}}, \tag{144}$$

where for $(1 - z)^{\frac{1}{3}}$ the principal value is taken and ϵ is a third root of unit. Let x', y' be a second solution of (143) for which the root of unit ϵ in (144) has the same value and let ξ', η', z' be the associated values of ξ, η, z. Setting δ for the number

$$(1 - z)^{\frac{1}{3}} A_g(z) - (1 - z')^{\frac{1}{3}} B_g(z) = \epsilon^{-1}\left(\frac{\eta}{\xi} A_g(z) - \frac{\eta'}{\xi'} B_g(z)\right),$$

then δ^3 is a number from $\mathfrak{K}$ because $(\eta/\xi)^3$ and the quotients $\xi : \xi', \eta : \eta'$ are elements of $\mathfrak{K}$. Furthermore, the number $h\xi'\xi^{3n+1-g}\delta$ is integral; if this number is $\neq 0$, then its absolute value is at least 1. One now chooses $g = 0$ or $= 1$ so that $\delta \neq 0$; then

$$|\delta| \geq \frac{1}{h\left|\xi'\xi^{3n+1-g}\right|} = \frac{\left|z'\right|^{\frac{1}{3}} |z|^{n+\frac{1-g}{3}}}{h\left|3k\sqrt{-3d}\right|^{n+\frac{2-g}{3}}}. \tag{145}$$

Since $|1 - z| = 1$, hence z neither lies near 1 nor near ∞, (139) is applicable. Taking into account the inequality $h \leq \gamma_1^n$ and (145), it follows that

$$|z|^{\frac{1}{3}} |z|^{n+\frac{1-g}{3}} < \gamma_2^n (k\sqrt{d})^{n+\frac{2-g}{3}} \left(|z|^{2n+1-g} + \left|z'\right|\right) \tag{146}$$

with constant γ_2.[23]

---

[23] Footnote by the Editors: This should be $|z'|^{1/3}|$ instead of $|z|^{1/3}$. See "Berichtigungen und Bemerkungen" on p. 339 in [C.L. Siegel, Gesammelte Abhandlungen. Band IV. Edited by K. Chandrasekharan and H. Maass. With corrections to the first three volumes. Springer-Verlag, Berlin-New York, 1979] for this correction.

If $\epsilon_1, \epsilon_2, \epsilon_3$ are the three third roots of unit, then it holds

$$(\epsilon_1\xi - \eta)(\epsilon_2\xi - \eta)(\epsilon_3\xi - \eta) = 3k\sqrt{-3d}.$$

Among the three factors of the left-hand side let $(\epsilon_1\xi - \eta)$ be the smallest in absolute value. Since $\eta : \xi$ has absolute value 1, it follows that $|\epsilon_2\xi - \eta| \geq |\xi|, |\epsilon_3\xi - \eta| \geq |\xi|$, and therefore

$$|\epsilon_1\xi - \eta| \leq \frac{\left|3k\sqrt{-3d}\right|}{|\xi|^2}.$$

If now $|z| < 1$, then clearly $\left|1 - (1-z)^{\frac{1}{3}}\right| < 1$, hence by using (144) also $|\epsilon\xi - \eta| < |\xi|$, and consequently $\epsilon_1 = \epsilon$. But if $|z| \geq 1$, then because of $|\eta| = |\xi|$ the inequality

$$|\epsilon\xi - \eta| \leq 2\,|\xi| \leq 2\,|\xi z| = \frac{\left|6k\sqrt{-3d}\right|}{|\xi|^2}$$

certainly holds. Therefore in any case

$$|\epsilon\xi - \eta| \leq \frac{\left|6k\sqrt{-3d}\right|}{|\xi|^2}$$

and also

$$\left|\epsilon\xi' - \eta'\right| \leq \frac{\left|6k\sqrt{-3d}\right|}{|\xi'|^2},$$

hence

(147) $$\left|\xi\eta' - \eta\xi'\right| \leq 6k\left|\sqrt{-3d}\right|\left(\frac{|\xi'|}{|\xi|^2} + \frac{|\xi|}{|\xi'|^2}\right).$$

The quadratic form $\xi\eta$ has the discriminant $-3d$; this gives

$$\xi\eta' - \eta\xi' = (xy' - yx')\sqrt{-3d}.$$

If now the two solutions x, y and x', y' of (143) are not identical, then $xy' - yx' \neq 0$, and therefore $\left|\xi\eta' - \eta\xi'\right| \geq \left|\sqrt{-3d}\right|$. If $\left|\xi'\right| \geq |\xi|$, then from (147) it follows that

$$|\xi|^2 \leq 12k\left|\xi'\right|$$

$$|z|^2 = \frac{\left|3k\sqrt{-3d}\right|^2}{|\xi|^6} \geq \frac{\left|3k\sqrt{-3d}\right|^2}{(12k)^3\,|\xi'|^3} = \frac{\left|3k\sqrt{-3d}\right|}{(12k)^3}\left|z'\right|$$

(148) $$\left|\frac{\gamma_3 k^2}{\sqrt{d}}z\right|^2 \geq \left|\frac{\gamma_3 k^2}{\sqrt{d}}z'\right|.$$

One now arranges all the solutions of (143) for which the root of unit ϵ in (144) has the same value, by growing values of $|\xi|$, say $|\xi_0| \le |\xi_1| \le |\xi_2| \le \cdots$. For the corresponding values $z_0, z_1, z_2, \ldots$ of z one has by (148)

$$\left|\frac{\gamma_3 k^2}{\sqrt{d}} z_{r+s}\right| \le \left|\frac{\gamma_3 k^2}{\sqrt{d}} z_r\right|^{2^s}. \qquad (r = 0, 1, \ldots; s = 1, 2, \ldots)$$

If at least seven solutions corresponding to ϵ exist, then one especially chooses $z = z_3, z' = z_6$. Since $|z_0| \le 2$, it follows that

$$|z| < \gamma_4 \left(\frac{k^2}{\sqrt{d}}\right)^7. \tag{149}$$

If one assumes that

$$d \ge \gamma_3^2 k^4, \tag{150}$$

then the exponent ν in the equation

$$|z'| = |z|^\nu$$

is ≥ 8. Using (146), then

$$\gamma_2^n (k\sqrt{d})^{n+\frac{2-g}{3}} (|z|^{n+\frac{2}{3}(1-g)-\frac{\nu}{3}} + |z|^{\frac{2}{3}\nu-n-\frac{1}{3}(1-g)}) > 1.$$

Herein one sets $n = \left[\frac{\nu+g}{2}\right]$,[24] then it is

$$\begin{aligned} n + \frac{2}{3}(1-g) - \frac{1}{3}\nu &\ge \frac{1}{6}\nu - \frac{1}{2} > 0 \\ \frac{2}{3}\nu - n - \frac{1}{3}(1-g) &\ge \frac{1}{6}\nu - \frac{1}{2} > 0 \\ n + \frac{1}{3}(2-g) &\le \frac{1}{2}\nu + \frac{5}{6}. \end{aligned}$$

This gives, together with (149), the estimate

$$\gamma_5^\nu (k\sqrt{d})^{\frac{1}{2}\nu+\frac{5}{6}} \left(\frac{k^2}{\sqrt{d}}\right)^{\frac{7}{6}\nu-\frac{7}{2}} > 1,$$

[24]Footnote by the Editors: Siegel himself pointed out in [C.L. Siegel, Einige Erläuterungen zu Thues Untersuchungen über Annäherungswerte algebraischer Zahlen und diophantische Gleichungen, Nachr. Akad. Wiss. Göttingen Math.-Phys. Kl. II 1970, 169–195] that one has to take $n = \left[\frac{v}{2} + \frac{13}{57}\right]$. Siegel goes on to say that with this choice of v the correspondingly changed formulae below lead to the exponent -62 instead of -33 and that the results of the newer paper show that the original assertion is still correct. See also "Berichtigungen und Bemerkungen" on p. 339 in [C.L. Siegel, Gesammelte Abhandlungen. Band IV. Edited by K. Chandrasekharan and H. Maass. With corrections to the first three volumes. Springer-Verlag, Berlin-New York, 1979].

and also

$$\gamma_6(k\sqrt{d})^{\frac{1}{2}\cdot 8+\frac{5}{6}}\left(\frac{k^2}{\sqrt{d}}\right)^{\frac{7}{6}\cdot 8-\frac{7}{2}} > 1.$$

This inequality leads to a contradiction if

$$d > \gamma_7 k^{33}. \tag{151}$$

If d is so large that (150) and (151) are satisfied, then there exist at most 6 solutions which correspond to this ϵ and in total at most 18 solutions to $\phi(x, y) = k$. One can mention that the bound 18 can be reduced a little by using some clever tricks; however, details will not be given here.

For the finitely many positive discriminants that do not satisfy (150) and (151) only finitely many classes of non-equivalent cubic forms exist. Therefore it is proved that the number of solutions of $\phi(x, y) = k$, for arbitrary d, lies below a bound which only depends on k.

This consideration can also be applied to cubic forms with negative discriminant d, however for small values of $|d|$ one does not obtain the sharp results of DELAUNAY and NAGELL.

Even easier is the study of the diophantine equation

$$ax^n - by^n = k$$

for arbitrary but fixed $n \geq 3$. It is possible to prove that it has at most one solution in natural numbers x, y provided that $|ab|$ lies above a bound which only depends on k and n. For $n = 3$ and $k = 1$ or 3 this assertion is contained in a precise theorem by NAGELL.

16.

Über einige Anwendungen diophantischer Approximationen

Abhandlungen der Preußischen Akademie der Wissenschaften.
Physikalisch-mathematische Klasse 1929, Nr. 1

Die bekannte einfache Schlußweise, daß bei einer Verteilung von mehr als n Dingen auf n Fächer in mindestens einem Fach mindestens zwei Dinge gelegen sind, enthält eine Verallgemeinerung des euklidischen Algorithmus, welche sich durch die Untersuchungen von Dirichlet, Hermite und Minkowski als die Quelle wichtiger arithmetischer Gesetze erwiesen hat. Sie liefert speziell eine Aussage darüber, wie genau sich *mindestens* die Zahl 0 durch eine lineare Verbindung

$$L = h_0 \omega_0 + \cdots + h_r \omega_r$$

aus geeigneten ganzen rationalen Zahlen $h_0, \cdots, h_r$, die, absolut genommen, höchstens gleich einer gegebenen natürlichen Zahl H und nicht sämtlich gleich 0 sein sollen, und gegebenen Zahlen $\omega_0, \cdots, \omega_r$ approximieren läßt; und zwar gilt für die beste Annäherung sicherlich

$$|L| \leq (|\omega_0| + \cdots + |\omega_r|) H^{-r},$$

also eine Aussage, die von dem feineren arithmetischen Verhalten der Zahlen $\omega_0, \cdots, \omega_r$ nicht abhängt.

Der Ausdruck L werde als Näherungsform bezeichnet. Fragt man nun danach, wie genau sich *höchstens* die Zahl 0 durch die Näherungsform $h_0 \omega_0 + \cdots + h_r \omega_r$ approximieren lasse, so hängt offenbar jede nichttriviale Antwort durchaus von den arithmetischen Eigenschaften der gegebenen Zahlen $\omega_0, \cdots, \omega_r$ ab.

In dieser Frage ist insbesondere das Problem enthalten, zu untersuchen, ob eine gegebene Zahl ω transzendent ist; man hat ja nur $\omega_0 = 1$, $\omega_1 = \omega, \cdots, \omega_r = \omega^r$, $H = 1, 2, 3, \cdots$, $r = 1, 2, 3, \cdots$ zu wählen. Durch die Forderung, sogar eine von 0 verschiedene untere Schranke für $|L|$ als Funktion von H und r anzugeben, bekommt das Transzendenzproblem eine positive Wendung.

Auch die obere Abschätzung der Anzahl der auf einer algebraischen Kurve gelegenen Gitterpunkte, also speziell die Untersuchung der Endlichkeit dieser Anzahl, führt, wie sich später zeigen wird, auf die Bestimmung einer positiven unteren Schranke für den absoluten Wert einer gewissen Näherungsform.

Analog zu den arithmetischen Problemen der oberen und unteren Abschätzung von $|L|$ ist eine algebraische Fragestellung. Es mögen nunmehr $\omega_0(x), \cdots, \omega_r(x)$ Reihen nach Potenzen einer Variabeln x bedeuten, und es seien die Polynome $h_0(x), \cdots, h_r(x)$ höchstens vom Grade H, nicht sämtlich identisch gleich 0 und außerdem so beschaffen, daß die Potenzreihe für die Näherungsform

$$L(x) = h_0(x)\,\omega_0(x) + \cdots + h_r(x)\,\omega_r(x)$$

mit einer möglichst hohen Potenz von x beginnt; gesucht wird eine untere und eine obere Abschätzung des Exponenten dieser Potenz von x. Das algebraische Problem ist von einfacherer Art als das arithmetische; es führt auf die Bestimmung des Ranges eines Systems linearer Gleichungen.

Zwischen beiden Problemen, dem algebraischen und dem arithmetischen, kommt ein Zusammenhang dadurch zustande, daß man für x einen speziellen rationalen Zahlwert ξ aus dem gemeinsamen Konvergenzbereich der Potenzreihen wählt und die Koeffizienten dieser Potenzreihen als rationale Zahlen voraussetzt. Dann können nämlich auch die

Koeffizienten der Polynome $h_0(x), \cdots, h_r(x)$ rationalzahlig gewählt werden; und durch Multiplikation mit dem Hauptnenner der rationalen Zahlen $h_0(\xi), \cdots, h_r(\xi)$ geht nunmehr die algebraische Näherungsform $L(x)$ in eine arithmetische über, falls nicht sämtliche Zahlen $h_0(\xi), \cdots, h_r(\xi)$ gleich 0 sind. Doch entsteht naturgemäß im allgemeinen aus der besten algebraischen Approximation nicht die beste arithmetische Approximation.

Für die untere Abschätzung des Ausdrucks $|h_0\omega_0 + \cdots + h_r\omega_r|$ unter den Bedingungen $|h_0| \leq H, \cdots, |h_r| \leq H$ bietet sich folgende Möglichkeit:

Die Zahlen $\omega_0, \cdots, \omega_r$ seien nicht sämtlich gleich 0. Es sollen $r+1$ Näherungsformen

$$L_k = h_{k0}\omega_0 + \cdots + h_{kr}\omega_r \qquad (k = 0, \cdots, r)$$

betrachtet werden, deren Koeffizienten ganz rational und absolut genommen $\leq H$ sind. Der Wert der Determinante $|h_{kl}|$ sei von 0 verschieden, und das Maximum der $r+1$ Zahlen $|L_k|$ werde mit M bezeichnet. Es sei L eine weitere Näherungsform und h der größte unter den absoluten Beträgen ihrer Koeffizienten. Da die $r+1$ Formen $L_0, \cdots, L_r$ linear unabhängig sind, so können unter ihnen r gewisse, etwa $L_1, \cdots, L_r$, ausgewählt werden, welche von L linear unabhängig sind. Es sei (λ_{kl}) reziprok zur Matrix der Koeffizienten von $L, L_1, \cdots, L_r$; dann gilt die Abschätzung

$$|\lambda_{k0}| \leq r!\,H^r \qquad (k = 0, \cdots, r)$$

$$|\lambda_{kl}| \leq r!\,hH^{r-1} \qquad (k = 0, \cdots, r;\ l = 1, \cdots, r)$$

für die absoluten Beträge der Elemente λ_{kl}. Aus den Gleichungen

(1) $$\omega_k = \lambda_{k0}L + \lambda_{k1}L_1 + \cdots + \lambda_{kr}L_r \qquad (k = 0, \cdots, r)$$

folgt daher

(2) $$|L| \geq \frac{|\omega_k|}{r!\,H^r} - \frac{rMh}{H}.$$

Wenn nun mit wachsendem H die von H abhängige Zahl M stärker 0 wird als H^{1-r}, so liefert (2) eine positive untere Schranke für $|L|$. Diese Bedingung ist also hinreichend für lineare Unabhängigkeit der Größen $\omega_0, \cdots, \omega_r$ im Körper der rationalen Zahlen.

Ein entsprechendes Kriterium gilt für lineare Unabhängigkeit von Potenzreihen im Körper der rationalen Funktionen. Es seien nämlich $\omega_0(x), \cdots, \omega_r(x)$ Potenzreihen, die nicht sämtlich identisch verschwinden; es seien

$$L_k(x) = h_{k0}(x)\omega_0(x) + \cdots + h_{kr}(x)\omega_r(x), \qquad (k = 0, \cdots, r)$$

$r+1$ Näherungsformen, deren Koeffizienten $h_{kl}(x)$ Polynome vom Grade H sind, und M der kleinste Exponent, der in den Potenzreihen für $L_0, \cdots, L_r$ wirklich auftritt. Die Determinante $|h_{kl}(x)|$ sei nicht identisch gleich 0. Es sei $L(x)$ eine weitere Näherungsform mit Koeffizienten vom Grade h. Die kleinsten Exponenten in den Potenzreihen $L(x)$ und $\omega_k(x)\,(k = 0, \cdots, r)$ seien μ und μ_k. Beachtet man nun, daß die Determinante $|h_{kl}(x)|$ den Grad $rH+h$ besitzt, so folgt aus einer zu (1) analogen Gleichung die Abschätzung

(3) $$rH + h + \mu_k \geq \min(\mu, M). \qquad (k = 0, \cdots, r)$$

Wenn nun die Differenz $M - rH$ mit wachsendem H unendlich wird, so folgt aus (3) eine obere Schranke für μ. Dies ist insbesondere hinreichend für lineare Unabhängigkeit der Potenzreihen $\omega_0(x), \cdots, \omega_r(x)$ im Körper der rationalen Funktionen.

Bei der Anwendung dieses Kriteriums liegt die Schwierigkeit in der Forderung des Nichtverschwindens der Determinante $|h_{kl}(x)|$. Um zu Fällen zu gelangen, bei denen diese Schwierigkeit überwunden werden kann, sei fortan vorausgesetzt, daß jede der Funktionen $\frac{d\omega_0(x)}{dx}, \cdots, \frac{d\omega_r(x)}{dx}$ sich homogen linear durch die Funktionen $\omega_0(x), \cdots, \omega_r(x)$ selber ausdrücken lasse, und zwar mit Koeffizienten, die rationale Funktionen von x sind. Es gilt dann also ein System von homogenen linearen Differentialgleichungen erster Ordnung

(4) $$\frac{d\omega_k}{dx} = a_{k0}\omega_0 + \cdots + a_{kr}\omega_r; \qquad (k = 0, \cdots, r)$$

und es entsteht durch Differentiation einer Näherungsform $L(x)$ wieder eine solche, wenn noch mit dem Polynom multipliziert wird, welches als Hauptnenner der Koeffizienten von $w_0, \cdots, w_r$ auftritt. Wird dies r-mal ausgeführt, so hat man insgesamt $r+1$ Näherungsformen. Es kann nun aber eintreten, daß die Determinante dieses Systems von $r+1$ Näherungsformen identisch gleich o ist; dann verschwindet auch die Determinante $\Delta(x)$ des Systems der $r+1$ linearen Formen $L, \frac{dL}{dx}, \cdots, \frac{d^rL}{dx^r}$, und umgekehrt. Die Bedeutung des identischen Verschwindens von Δ ergibt sich aus folgendem *Hilfssatz*:

Es sei

$$w_k = c_0 w_{k0} + \cdots + c_r w_{kr}, \qquad (k = 0, \cdots, r)$$

wo $c_0, \cdots, c_r$ *beliebige Konstanten bedeuten, die allgemeine Lösung des Systems* (4). *Die Determinante des Systems der* $r+1$ *linearen Formen* $L(x) = h_0(x) w_0(x) + \cdots + h_r(x) w_r(x)$, $\frac{dL}{dx}, \cdots, \frac{d^rL}{dx^r}$ *von* $w_0, \cdots, w_r$ *verschwindet dann und nur dann identisch, wenn zwischen den* $r+1$ *Funktionen*

$$f_l = h_0 w_{0l} + \cdots + h_r w_{rl} \qquad (l = 0, \cdots, r)$$

eine homogene lineare Gleichung mit konstanten Koeffizienten besteht.

Ist nämlich

$$\frac{d^kL}{dx^k} = b_{k0} w_0 + \cdots + b_{kr} w_r \qquad (k = 0, \cdots, r)$$

mit $b_{0l} = h_l$, $b_{k+1,l} = \frac{db_{kl}}{dx} + b_{k0} a_{0l} + \cdots + b_{kr} a_{rl}$ für $k = 0, \cdots, r-1$ und $l = 0, \cdots, r$, so folgt unter der Annahme des identischen Verschwindens der Determinante $|b_{kl}| = \Delta$ eine Gleichung

$$g_0 \frac{d^sL}{dx^s} + g_1 \frac{d^{s-1}L}{dx^{s-1}} + \cdots + g_s L = 0,$$

wo $s \leq r$ ist und $g_0, \ldots, g_s$ gewisse Unterdeterminanten von $\Delta(x)$ bedeuten, von denen g_0 nicht identisch verschwindet. Dieser linearen Differentialgleichung s-ter Ordnung genügt die Funktion

$$L = \sum_{k=0}^{r} h_k w_k = \sum_{k=0}^{r} h_k \sum_{l=0}^{r} c_l w_{kl} = \sum_{l=0}^{r} c_l f_l,$$

also jede der $r+1$ Funktionen $f_0, \cdots, f_r$; und da ihre Anzahl größer als s ist, so besteht zwischen ihnen eine homogene lineare Gleichung mit konstanten Koeffizienten. Umgekehrt folgt aus einer solchen Gleichung durch r-malige Differentiation, daß die Determinante $\left|\frac{d^kf_l}{dx^k}\right|$ identisch gleich o ist, und aus der Matrizenrelation

$$\left(\frac{d^kf_l}{dx^k}\right) = (b_{kl})(w_{kl})$$

erhält man die Gleichung $|b_{kl}| = \Delta = 0$, wenn man beachtet, daß die Werte der Lösungen $w_{0l}, \cdots, w_{rl} (l = 0, \cdots, r)$ in einem regulären Punkt so vorgeschrieben werden können, daß die Determinante $|w_{kl}|$ in diesem Punkte nicht o ist und daher erst recht nicht identisch verschwindet.

Ein Beispiel liefert die Annahme $w_k = e^{kx}$, also $w_{kl} = e^{kx} e_{kl}$, wo (e_{kl}) die Einheitsmatrix bedeutet. Es wird $f_l = h_l(x) e^{lx}$, und es besteht zwischen $f_0, \cdots, f_r$ keine homogene lineare Gleichung mit konstanten Koeffizienten, weil e^x keine algebraische Funktion ist. Also verschwindet in diesem Falle Δ nicht identisch.

Es werde nun wieder vorausgesetzt, daß die Potenzreihen $w_0(x), \cdots, w_r(x)$ lauter rationale Koeffizienten besitzen und daß ξ eine rationale Zahl sei. Dann gewinnt man aus dem System der algebraischen Näherungsformen für die Funktionen $w_0(x), \ldots, w_r(x)$ ein System von arithmetischen Näherungsformen für die Zahlen $w_0(\xi), \ldots, w_r(\xi)$. Der wichtigste Punkt der ganzen Untersuchung ist nun die Konstruktion von Näherungsformen, für welche die Zahl $\Delta(\xi) \neq 0$ ist. Hierzu dient folgende Überlegung, die in einem speziellen Fall zuerst von Thue verwendet worden ist.

Die Potenzreihe für die Näherungsform L enthalte als kleinsten Exponenten die Zahl γ. Durch Multiplikation mit dem Hauptnenner $N(x)$ der rationalen Funktionen a_{kl} aus (4) erhält man aus $\frac{dL}{dx}$ eine Näherungsform L_1, deren Potenzreihe keine kleinere als die $(\gamma-1)$-te Potenz von x enthält. Man bilde nun die Determinante $D(x)$ der $r+1$ Näherungsformen L, $N\frac{dL}{dx}=L_1$, $N\frac{dL_1}{dx}=L_2$, $\cdots$, $N\frac{dL_{r-1}}{dx}=L_r$. Ist ν eine obere Schranke für den Grad der $(r+1)^2+1$ Polynome N, Na_{kl}, und sind die Koeffizienten der Form L vom Grade H, so ist $D(x)$ vom Grade $H+(H+\nu)+\cdots+(H+r\nu)=(r+1)H+\frac{r(r+1)}{2}\nu$. Andererseits läßt sich $D(x)\omega_k(x)$ linear homogen durch L, L_1, $\cdots$, L_r ausdrücken, etwa

$$D\omega_k=\Lambda_{k0}L+\Lambda_{k1}L_1+\cdots+\Lambda_{kr}L_r, \qquad (k=0,\cdots,r) \tag{5}$$

und zwar mit Koeffizienten Λ_{kl}, die Polynome in x sind. Die Funktion $D(x)\omega_k(x)$ verschwindet daher für $x=0$ mindestens von der Ordnung $\gamma-r$. Setzt man voraus, daß nicht alle Potenzreihen ω_0, $\cdots$, ω_r den Faktor x haben, den man ja sonst durch Division beseitigen könnte, so folgt, daß $D(x)$ für einen von 0 verschiedenen Wert $x=\xi$ höchstens von der Ordnung $(r+1)H+\frac{r(r+1)}{2}\nu+r-\gamma$ verschwindet, falls $D(x)$ nicht identisch gleich 0 ist. Nun kann man aber durch geeignete Wahl von L erreichen, daß $\gamma\geq(r+1)H+r$ ist; die $r+1$ Polynome $h_0(x)$, $\cdots$, $h_r(x)$ vom Grade H enthalten nämlich $(r+1)(H+1)$ Zahlenkoeffizienten, welche γ homogenen linearen Gleichungen genügen müssen. Dann verschwindet aber $D(x)$ für $x=\xi$ von einer Ordnung s, die unterhalb der von H freien Schranke $\frac{r(r+1)}{2}\nu$ gelegen ist, und die s-te Ableitung von $D(x)$ ist für $x=\xi$ von 0 verschieden. Ferner gilt die Gleichung (5) *identisch* in ω_0, $\cdots$, ω_r; differentiiert man sie s-mal und benutzt dabei (4) zur Elimination der Ableitungen von ω_0, $\cdots$, ω_r, so gilt auch die so entstehende Gleichung identisch in ω_0, $\cdots$, ω_r. Setzt man noch $N\frac{dL_r}{dx}=L_{r+1}$, $\cdots$, $N\frac{dL_{r+s-1}}{dx}=L_{r+s}$, so ist nach (5) der Ausdruck $N^s(\xi)D^{(s)}(\xi)\omega_k$ identisch in ω_0, $\cdots$, ω_r eine homogene lineare Verbindung von $L(\xi)$, $L_1(\xi)$, $\cdots$, $L_{r+s}(\xi)$. Nimmt man nun noch an, daß $N(\xi)\neq 0$ ist, daß also ξ von den singulären Stellen des Systems (4) verschieden ist, so erhält man ω_0, $\cdots$, ω_r als lineare Verbindung von $L(\xi)$, $L_1(\xi)$, $\cdots$, $L_{r+s}(\xi)$; unter den $r+s+1$ Formen $L(\xi)$, $\cdots$, $L_{r+s}(\xi)$ gibt es demnach $r+1$ linear unabhängige. Damit hat man $r+1$ arithmetische Näherungsformen für die Zahlen $\omega_0(\xi)$, $\cdots$, $\omega_r(\xi)$ gefunden, deren Determinante $\neq 0$ ist.

Für die zahlentheoretische Anwendung ist es nun aber noch erforderlich, daß die soeben konstruierten Näherungsformen eine in dem früher präzisierten Sinne günstige arithmetische Approximation liefern, daß also die Koeffizienten von $L(x)$, $\cdots$, $L_{r+s}(x)$ nicht »allzu große« ganze rationale Zahlen enthalten. Da die Zahl s unterhalb einer von H freien Schranke gelegen ist, so handelt es sich im wesentlichen noch um eine günstige Abschätzung der Koeffizienten der Polynome $h_0(x)$, $\cdots$, $h_r(x)$ in $L=h_0\omega_0+\cdots+h_r\omega_r$.

Dies läßt sich bei dem erwähnten Beispiele $\omega_k(x)=e^{kx}$ leicht ausführen, da sich die Koeffizienten explizit durch r und H ausdrücken lassen; so erhält man den Transzendenzbeweis für e in der ersten von Hermite gegebenen Fassung und zugleich eine positive untere Schranke für den Abstand einer beliebigen algebraischen Zahl von e. Es sei noch darauf hingewiesen, daß sich aus Hermites Formeln ohne weiteres die Transzendenz von π und sogar eine positive untere Schranke für den Abstand einer beliebigen algebraischen Zahl von π ergibt, wenn noch beachtet wird, daß die Norm einer von 0 verschiedenen ganzen algebraischen Zahl absolut genommen ≥ 1 ist.

Ein anderes Beispiel wird, aber nur für den Fall $r=1$, durch die bekannten Kettenbrüche für die Quotienten von hypergeometrischen Reihen geliefert. Insbesondere ist der Kettenbruch der Funktion $(1-x)^\alpha$ von Thue benutzt worden, um die Approximation der

Wurzeln natürlicher Zahlen durch rationale Zahlen zu untersuchen, und dies war der Ausgangspunkt für die Entdeckung des THUEschen Satzes über diophantische Gleichungen.

In andern Fällen aber gelingt es nicht, für die Zahlenkoeffizienten in der algebraisch günstigen Näherungsform, deren Potenzreihe ja durch $x^{(r+1)(H+1)-1}$ teilbar ist, eine über das Triviale hinausgehende Abschätzung zu finden, und die triviale Schranke reicht, wie man leicht sieht, für die Anwendung des arithmetischen Kriteriums nicht aus. Da muß nun der Gedankengang, welcher zu einem System von arithmetischen Näherungsformen für $\omega_0(\xi), \cdots, \omega_r(\xi)$ mit nicht verschwindender Determinante geführt hat, etwas modifiziert werden. Man hat zwischen den beiden Forderungen der guten algebraischen und arithmetischen Annäherung an 0 ein Kompromiß zu schließen, indem man für die Zahl γ einen kleineren als den früher angegebenen Wert zuläßt, für die Zahl s also einen größeren, als Gewinn aber bessere Schranken für die Koeffizienten von $h_0(x), \cdots, h_r(x)$ erhält. Diese Idee ist ebenfalls zuerst von THUE angewendet worden. Die Abschätzung der Koeffizienten ergibt sich durch Benutzung der eingangs erwähnten DIRICHLETschen Schlußweise und soll in Form eines *Hilfssatzes* ausgesprochen werden:

Es seien

$$y_1 = a_{11}x_1 + \cdots + a_{1n}x_n$$
$$\cdots\cdots\cdots$$
$$y_m = a_{m1}x_1 + \cdots + a_{mn}x_n$$

m lineare Formen in n Variabeln mit ganzen rationalen Koeffizienten. Es sei $n > m$. Die absoluten Beträge der mn Koeffizienten a_{kl} seien sämtlich nicht größer als eine natürliche Zahl A. Dann sind die homogenen linearen Gleichungen $y_1 = 0, \cdots, y_m = 0$ lösbar in ganzen rationalen Zahlen $x_1, \cdots, x_n$, welche nicht sämtlich 0, aber, absolut genommen, kleiner als $1 + (nA)^{\frac{m}{n-m}}$ sind.

Zum Beweise lasse man jede der Variabeln $x_1, \cdots, x_n$ unabhängig voneinander die Werte $0, \pm 1, \pm 2, \cdots, \pm H$ durchlaufen; man erhält insgesamt $(2H+1)^n$ Gitterpunkte im Raum mit den rechtwinkligen kartesischen Koordinaten $y_1, \cdots, y_m$, welche aber nicht alle verschieden zu sein brauchen. Jede Koordinate jedes dieser Gitterpunkte liegt zwischen den Werten $-nAH$ und $+nAH$. Es gibt genau $(2nAH+1)^m$ verschiedene Gitterpunkte im m-dimensionalen Raum, deren Koordinaten zwischen $-nAH$ und $+nAH$ gelegen sind. Ist nun

$$(6) \qquad (2nAH+1)^m < (2H+1)^n,$$

so koinzidieren zwei zu verschiedenen Systemen $x_1, \cdots, x_n$ gehörende Gitterpunkte $y_1, \cdots, y_m$; und durch Subtraktion dieser beiden Systeme erhält man eine Lösung von $y_1 = 0, \cdots, y_m = 0$ in ganzen rationalen Zahlen $x_1, \cdots, x_n$, die nicht sämtlich 0 und absolut $\leq 2H$ sind. Die Bedingung (6) ist aber erfüllt, wenn für $2H$ die gerade ganze Zahl des Intervalls

$$(nA)^{\frac{m}{n-m}} - 1 \leq 2H < (nA)^{\frac{m}{n-m}} + 1$$

gewählt wird.

Die im vorhergehenden skizzierte Methode zur Auffindung einer positiven unteren Schranke für den Ausdruck $|h_0\omega_0 + \cdots + h_r\omega_r|$ wird in dieser Abhandlung auf zwei verschiedene Probleme angewendet werden. Der erste Teil behandelt vorwiegend den Nachweis der Transzendenz des Wertes der Zylinderfunktion $J_0(x)$ für jedes algebraische von 0 verschiedene Argument. Der zweite Teil beschäftigt sich mit der Aufgabe, alle algebraischen Kurven zu finden, die durch unendlich viele Gitterpunkte der Ebene oder allgemeiner des n-dimensionalen Raumes hindurchgehen; es wird gezeigt, daß dies nur bei Geraden und Hyperbeln eintreten kann sowie bei gewissen andern Kurven, die sich hieraus durch eine einfache Transformation ergeben und ebenfalls das Geschlecht 0 besitzen.

Den Anlaß zur Beschäftigung mit den im ersten Teil zu besprechenden Problemen gaben die schönen Irrationalitätsuntersuchungen von W. MAIER. Der zweite Teil verdankt seinen Ursprung den wichtigen Resultaten über die arithmetischen Eigenschaften der algebraischen Kurven, welche A. WEIL entdeckt und unlängst in seiner These veröffentlicht hat.

Erster Teil: Über transzendente Zahlen.

MAX DEHN gewidmet.

Durch die Sätze von HERMITE und LINDEMANN ist die Frage nach den arithmetischen Eigenschaften der Werte der Exponentialfunktion für algebraisches Argument beantwortet worden. Während durch das Additionstheorem der Exponentialfunktion jede algebraische Gleichung zwischen Werten dieser Funktion auf eine lineare Gleichung reduziert wird, so gilt etwas Entsprechendes bei andern Funktionen nicht mehr; und darin liegt die Schwierigkeit einer Verallgemeinerung der Überlegungen von HERMITE. Für keine weitere der Funktionen, welche für die Analysis Bedeutung haben, ist bisher ein Satz von analogem Umfang bewiesen worden, wie er für die Exponentialfunktion gilt.

Für die Zylinderfunktion

$$J_0(x) = \sum_{n=0}^{\infty} \frac{(-1)^n}{n!\; n!} \left(\frac{x}{2}\right)^{2n}$$

sind von verschiedenen Autoren Irrationalitätsuntersuchungen ausgeführt worden. HURWITZ und STRIDSBERG haben nachgewiesen, daß $J_0(x)$ für jeden von 0 verschiedenen rationalen Wert von x^2 irrational ist, und MAIER hat, darüber hinausgehend, in höchst scharfsinniger Weise gezeigt, daß für diese x der Wert $J_0(x)$ noch nicht einmal eine quadratische Irrationalität ist.

Im folgenden soll bewiesen werden, daß $J_0(x)$ für jedes von 0 verschiedene algebraische x einen transzendenten Wert hat. Es wird sich zugleich das weitergehende Resultat ergeben, daß zwischen den Zahlen $J_0(x)$ und $J_0'(x)$ keine algebraische Gleichung mit rationalen Koeffizienten besteht, und zwar wird für den absoluten Betrag eines beliebigen Polynoms in $J_0(x)$ und $J_0'(x)$, dessen Koeffizienten rationale Zahlen sind, eine positive untere Schranke als Funktion dieser Koeffizienten angegeben werden. Allgemeiner wird das Analogon des LINDEMANNschen Satzes bewiesen, daß zwischen den Zahlen $J_0(\xi_1)$, $J_0'(\xi_1)$, $\cdots$, $J_0(\xi_k)$, $J_0'(\xi_k)$ keine algebraische Gleichung mit rationalen Koeffizienten besteht, falls ξ_1^2, $\cdots$, ξ_k^2 voneinander und von 0 verschiedene algebraische Zahlen bedeuten.

Der Beweis erfolgt mit der in der Einleitung auseinandergesetzten Methode. Für die Anwendung des ersten Hilfssatzes wird insbesondere der Satz benötigt, daß die Funktion $J_0(x)$ keiner algebraischen Differentialgleichung erster Ordnung genügt, deren Koeffizienten Polynome in x sind. Es erscheint also der oben ausgesprochene Satz, daß für numerisches algebraisches $x \neq 0$ zwischen den Zahlen $J_0(x)$, $J_0'(x)$ und x keine algebraische Gleichung mit algebraischen Koeffizienten besteht, als Konsequenz des Satzes, daß zwischen den Funktionen $J_0(x)$, $J_0'(x)$ und x keine algebraische Gleichung identisch in x besteht. Dies dürfte vielleicht darauf hindeuten, daß auch in allgemeineren Fällen numerische Relationen durch Spezialisierung von Funktionalgleichungen entstehen, so daß die Analysis in diesem Sinne die Arithmetik umfaßt.

§ 1. Funktionentheoretische Hilfsmittel.

In diesem Paragraphen soll untersucht werden, welche algebraischen Funktionalgleichungen zwischen den Lösungen einer BESSELschen Differentialgleichung

$$\frac{d^2y}{dx^2} + \frac{1}{x}\frac{dy}{dx} + \left(1 - \frac{\lambda^2}{x^2}\right) y = 0, \tag{7}$$

ihren Ableitungen und der unabhängigen Variabeln x bestehen. Es stellt sich heraus, daß außer den wohlbekannten Relationen keine weiteren existieren, oder daß vielmehr jede weitere aus diesen durch rationale Umformung hervorgeht.

Satz 1:

Jede der BESSELschen Differentialgleichung (7) *genügende algebraische Funktion y ist identisch gleich* 0.

Beweis: In der Umgebung von $x = \infty$ gilt eine Entwicklung

(8) $$y = a_0 x^r + a_1 x^s + \cdots$$

nach fallenden Potenzen von x mit ganzen oder gebrochenen Exponenten. Trägt man (8) in (7) ein, so folgt $a_0 = 0$, also das identische Verschwinden von y.

Satz 2:

Es sei y eine Lösung der Bessel*schen Differentialgleichung* (7). *Zwischen den Funktionen y, $\frac{dy}{dx}$ und x besteht dann und nur dann eine algebraische Gleichung mit konstanten Koeffizienten, wenn λ die Hälfte einer ungeraden Zahl ist.*

Beweis: Es bestehe die Gleichung

(9) $$P(y', y, x) = 0,$$

wo P ein irreduzibles Polynom der drei Argumente y', y, x bedeutet. Nach (7) genügt dann y auch der Differentialgleichung

(10) $$\frac{\partial P}{\partial y'}\left\{-\frac{1}{x}y' - \left(1 - \frac{\lambda^2}{x^2}\right) y\right\} + \frac{\partial P}{\partial y} y' + \frac{\partial P}{\partial x} = 0.$$

Dies ist wieder eine algebraische Gleichung zwischen y', y, x. Nach Satz 1 ist es unmöglich, die Variable y' aus den Gleichungen (9) und (10) zu eliminieren. Die mit x^2 multiplizierte linke Seite von (10) ist ein Polynom in y', y, x und muß daher durch das irreduzible Polynom P teilbar sein. Der Quotient ist, wie die Betrachtung des Grades sofort zeigt, ein quadratisches Polynom $ax^2 + bx + c$ in x allein. Die linke Seite von (10) ist gleich $\frac{dP}{dx}$, falls (7) erfüllt ist, also für *jede* Lösung der Besselschen Differentialgleichung. Da (7) homogen in y, y', y'' ist, so bleibt die Gleichung

$$\frac{dP}{dx} = \left(a + \frac{b}{x} + \frac{c}{x^2}\right) P$$

für jede Lösung von (7) richtig, wenn darin das Polynom P durch das Aggregat Q der Glieder höchster Dimension in y', y von P ersetzt wird. Durch Integration folgt

(11) $$Q(y', y, x) = k x^b e^{ax - \frac{c}{x}}$$

mit konstantem k. Nun seien y_1 und y_2 irgend zwei linear unabhängige Lösungen der Besselschen Differentialgleichung; dann genügt $y = \lambda_1 y_1 + \lambda_2 y_2$ für beliebige konstante Werte von λ_1 und λ_2 der Gleichung (11). Da Q ein homogenes, nicht konstantes Polynom in y', y ist, so ist die Integrationskonstante k in (11) ein homogenes Polynom in λ_1 und λ_2. Man wähle nun das Verhältnis $\lambda_1 : \lambda_2$ derart, daß $k = 0$ wird. Für die zugehörige Funktion y gilt dann die Gleichung

$$Q(y', y, x) = 0,$$

und folglich ist der Quotient $y' : y$ eine algebraische Funktion von x.

Gibt es also eine Lösung der Besselschen Differentialgleichung, die einer algebraischen Differentialgleichung erster Ordnung genügt, so gibt es auch eine solche Lösung, deren logarithmische Ableitung eine algebraische Funktion ist. Dieser Satz wird übrigens auf genau dieselbe Weise für Lösungen von beliebigen homogenen linearen Differentialgleichungen zweiter Ordnung mit algebraischen Koeffizienten bewiesen.

Die logarithmische Ableitung $z = y' : y$ jeder Lösung der Besselschen Differentialgleichung genügt der Riccatischen Gleichung

(12) $$\frac{dz}{dx} + z^2 + \frac{1}{x} z + 1 - \frac{\lambda^2}{x^2} = 0.$$

Ist nun z algebraisch, so gilt bei $x = \infty$ eine Entwicklung

$$z = a_0 x^{r_0} + a_1 x^{r_1} + \cdots \qquad (r_0 > r_1 > \cdots)$$

nach fallenden Potenzen von x mit ganzen oder gebrochenen Exponenten und von o verschiedenen Koeffizienten $a_0, a_1, \cdots$.

Folglich ist

$$\sum_{k=0}^{\infty} a_k (r_k + 1) x^{r_k - 1} + \sum_{k=0}^{\infty} \sum_{l=0}^{\infty} a_k a_l x^{r_k + r_l} + 1 - \frac{\lambda^2}{x^2} = 0.$$

Der Koeffizientenvergleich liefert zunächst $r_0 = 0$, $a_0^2 + 1 = 0$, $a_0 = \pm i$; ferner muß für $n = 1, 2, \cdots$ der Exponent r_n des Gliedes $a_0 a_n x^{r_0 + r_n}$ gleich einem der Exponenten -2, $r_k - 1$, $r_k + r_l$ $(k = 0, \cdots, n-1;\ l = 0, \cdots, n-1)$ sein; und daraus folgt die Ganzzahligkeit aller Exponenten $r_0, r_1, \cdots$. Speziell wird noch $r_1 = -1$, $2a_0 a_1 + a_0 = 0$, $a_1 = -\frac{1}{2}$.

Die Besselsche Differentialgleichung hat nun nur die singulären Punkte o und ∞; folglich kann auch die algebraische Funktion z nur in o und ∞ verzweigt sein. Soeben wurde bewiesen, daß jeder Zweig von z im Unendlichen regulär ist. Daher ist z eine rationale Funktion von x. Es sei

$$z = b x^s + \cdots$$

die Entwicklung bei $x = 0$; dann liefert (12) die Gleichung

$$b(s+1) x^{s-1} + b^2 x^{2s} - \frac{\lambda^2}{x^2} + \cdots = 0,$$

also $s = -1$ und $b = \pm \lambda$. Da jede von o und ∞ verschiedene Nullstelle von y nach (7) von erster Ordnung ist, so hat z in diesen Nullstellen von y Pole erster Ordnung mit dem Residuum 1. Folglich ist

$$z = \pm i \pm \frac{\lambda}{x} + \frac{1}{x - x_1} + \cdots + \frac{1}{x - x_k}$$

mit gewissen, von o verschiedenen Konstanten $x_1, \cdots, x_k$. Die Entwicklung bei $x = \infty$ liefert $a_1 = \pm \lambda + k$, also

$$\lambda = \pm (k + \tfrac{1}{2})$$

mit nicht negativem ganzen rationalen k.

Damit zwischen y', y, x eine algebraische Gleichung identisch in x bestehen kann, ist also notwendig, daß λ die Hälfte einer ungeraden Zahl ist. Dies ist aber auch hinreichend, denn bekanntlich genügen die beiden linear unabhängigen Funktionen

$$H_1 = (-1)^k \frac{(2x)^{k+\frac{1}{2}}}{\sqrt{\pi}} \frac{d^k}{d(x^2)^k} \frac{e^{ix}}{ix}$$

und

$$H_2 = (-1)^k \frac{(2x)^{k+\frac{1}{2}}}{\sqrt{\pi}} \frac{d^k}{d(x^2)^k} \frac{e^{-ix}}{-ix}$$

der Besselschen Differentialgleichung mit $\lambda = \pm (k + \frac{1}{2})$, und es hat jede homogene lineare Kombination von H_1 und H_2 offenbar die Form

$$y = \sqrt{x}\,(R(x) \cos x + S(x) \sin x),$$

wo $R(x)$ und $S(x)$ rationale Funktionen von x sind, so daß y einer Differentialgleichung erster Ordnung

$$P_1(x) y'^2 + P_2(x) y' y + P_3(x) y^2 = P_4(x)$$

genügt, deren Koeffizienten $P_1(x), \cdots, P_4(x)$ Polynome in x bedeuten.

Satz 3:

Es sei λ nicht die Hälfte einer ungeraden Zahl und es seien y_1, y_2 zwei linear unabhängige Lösungen der Besselschen Differentialgleichung (7). *Dann besteht zwischen den Funktionen y_1, $\frac{dy_1}{dx}$, y_2, x keine algebraische Gleichung mit konstanten Koeffizienten.*

Beweis: Es bestehe die Gleichung

$$P(y_1, y_1', y_2, x) = 0, \tag{13}$$

wo P ein Polynom der vier Argumente y_1, y_1', y_2, x bedeutet. Nach Satz 2 enthält P wirklich das Argument y_2. Es sei P irreduzibel und in y_2 vom genauen Grade $n \geq 1$. Der Koeffizient von y_2^n heiße $f(y_1, y_1', x)$.

Aus (13) folgt durch Differentiation nach x

$$\frac{dP}{dx} = \frac{df}{dx} y_2^n + n f y_2^{n-1} y_2' + \cdots = 0.$$

Nun gilt aber mit konstantem $\alpha \neq 0$

$$y_1 y_2' - y_2 y_1' = \frac{\alpha}{x} \tag{14}$$

und daher

$$\frac{dP}{dx} = \left(\frac{df}{dx} + n f \frac{y_1'}{y_1}\right) y_2^n + \cdots = 0. \tag{15}$$

Entfernt man hieraus noch y_1'' vermöge (7), so erhält man eine algebraische Gleichung zwischen y_1, y_1', y_2, x, welche in y_2 wieder vom Grade n ist. Nach Satz 2 ist aber die Elimination von y_2 aus dieser Gleichung und (13) unmöglich. Folglich unterscheidet sich das irreduzible Polynom P von der Funktion $\frac{dP}{dx}$, aus welcher y_2' und y_1'' mit Hilfe von (7) und (14) eliminiert sind, nur durch einen von y_2 freien Faktor, und zwar nach (15) offenbar durch den Faktor $f : \left(\frac{df}{dx} + n f \frac{y_1'}{y_1}\right)$. Demnach gilt

$$\frac{P'}{P} = \frac{f'}{f} + n \frac{y_1'}{y_1}$$

identisch in y_1, y_1', y_2, x, wenn nur (7) und (14) erfüllt sind. Man kann also in dieser Gleichung y_2 durch $y_2 + \lambda y_1$ mit beliebigem konstanten λ ersetzen, ohne daß sie ihre Gültigkeit verliert. Die Integration liefert

$$P(y_1, y_1', y_2 + \lambda y_1, x) = c(\lambda) f(y_1, y_1', x) y_1^n,$$

wo $c(\lambda)$ ein Polynom in λ mit konstanten Koeffizienten bedeutet. Wird nach λ differentiiert und $\lambda = 0$ gesetzt, so folgt

$$y_1 (n f y_2^{n-1} + \cdots) = c_1 f y_1^n$$

mit konstantem c_1. Dies ist eine Gleichung vom genauen Grade $n - 1$ für y_2. Da nun jede y_2 wirklich enthaltende Gleichung zwischen y_1, y_1', y_2, x in y_2 mindestens vom Grade n ist, so folgt $n = 1$.

Es gilt also

$$y_2 = g(y_1, y_1', x) : f(y_1, y_1', x), \tag{16}$$

wo g und f Polynome in y_1, y_1', x sind. Trägt man dies in (14) ein und ersetzt y_1'' mit Hilfe der Besselschen Differentialgleichung durch y_1' und y_1, so muß nach Satz 2 eine identische Gleichung in y_1', y_1, x entstehen. Insbesondere müssen sich die Glieder höchster Dimension in y_1 und y_1' bei (14) aufheben. Die rationale Funktion $g:f$ habe in y_1, y_1' die Dimension d. Wegen der Homogenität von (7) hat dann auch die Ableitung von $g:f$ dieselbe Dimension. Man behalte nun in $g:f$ nur die Glieder höchster Dimension im Zähler und im Nenner bei; dadurch geht y_2 über in eine homogene rationale Funktion z von y_1, y_1', deren Dimension gleich d ist, und die Dimension von $y_2 - z$ ist dann kleiner als d. Die rechte Seite von (14) hat die Dimension 0, folglich ist $d + 1 \geq 0$. Ist $d + 1 > 0$, so gilt nach (14)

$$y_1 z' - y_1' z = 0;$$

ist dagegen $d + 1 = 0$, so gilt

$$y_1 z' - y_1' z = \frac{\alpha}{x}.$$

Im ersten Falle ist $z = b y_1$ mit konstantem b, also $d = 1$; dann ersetze man in (16) die Funktion y_2 durch die Funktion $y_2 - b y_1$ und kommt so auf den zweiten Fall. In diesem ist z eine von y_1 linear unabhängige Lösung der Besselschen Differentialgleichung. Es gibt also eine rationale Funktion von y_1, y_1', x, die in y_1, y_1' homogen von der Dimension -1 sind und der Gleichung (7) genügt. Sie werde mit $R(y_1, y_1', x)$ bezeichnet. Nach Satz 2 genügt dann auch $R(\lambda_1 y_1 + \lambda_2 y_2, \lambda_1 y_1' + \lambda_2 y_2', x)$ der Besselschen Differentialgleichung, es gilt also

$$R(\lambda_1 y_1 + \lambda_2 y_2, \lambda_1 y_1' + \lambda_2 y_2', x) = \Lambda_1 y_1 + \Lambda_2 y_2, \tag{17}$$

wo Λ_1 und Λ_2 nur von λ_1 und λ_2 abhängen. Aus den Formeln

$$Ry_2' - \frac{dR}{dx} y_2 = \frac{\alpha}{x}\Lambda_1,$$

$$Ry_1' - \frac{dR}{dx} y_1 = -\frac{\alpha}{x}\Lambda_2$$

erkennt man, daß Λ_1 und Λ_2 homogene rationale Funktionen von λ_1, λ_2 mit der Dimension -1 sind. Demnach kann man das Verhältnis $\lambda_1 : \lambda_2$ so wählen, daß mindestens eine der Funktionen Λ_1, Λ_2 unendlich wird. Da y_1 und y_2 nicht proportional sind, so wird auch die rechte Seite von (17) unendlich. Mit den gefundenen Werten von λ_1 und λ_2 gilt nun für $y = \lambda_1 y_1 + \lambda_2 y_2$ die Gleichung

$$1 : R(y, y', x) = 0, \tag{18}$$

und das verstößt gegen Satz 2. Damit ist Satz 3 bewiesen.

Das zum Beweise der Sätze 1, 2, 3 benutzte Verfahren ist aus Ideen von Liouville und Riemann zusammengesetzt. Analog kann man bei beliebigen homogenen linearen Differentialgleichungen zweiter Ordnung mit algebraischen Koeffizienten verfahren, um alle algebraischen Relationen zwischen den Lösungen, ihren Ableitungen und der unabhängigen Variabeln aufzufinden.

Der elementare algebraische Gedankengang, welcher zum Beweise des Satzes 3 führte, ist vielleicht nicht ganz durchsichtig. Man kann nun den Satz noch auf eine wesentlich verschiedene Art ableiten, die zwar tiefe arithmetische Hilfsmittel benötigt, aber prinzipiell recht einfach ist. Es stellt sich nämlich heraus, daß zwischen y_1, y_1', y_2, x deswegen keine algebraische Relation bestehen kann, weil die Koeffizienten in der Potenzreihe für y_2 »zu verschieden« von denen in der Potenzreihe für y_1 sind. Der Beweis ist mit Untersuchungen verwandt, die Eisenstein und Tschebyscheff angestellt haben; seine Idee stammt von Maier. Zunächst soll als naheliegende Verallgemeinerung eines Eisensteinschen Satzes eine hinreichende Bedingung für algebraische Unabhängigkeit von Potenzreihen angegeben werden:

Es seien

$$f(x) = \sum_{n=0}^{\infty} \gamma_n x^n$$

$$f_\nu(x) = \sum_{n=0}^{\infty} \gamma_n^{(\nu)} x^n \qquad (\nu = 1, \cdots, h)$$

Potenzreihen, deren Koeffizienten sämtlich einem festen algebraischen Zahlkörper $\mathfrak{K}$ angehören. Gibt es nun zu jedem natürlichen r ein $n > r$ von der Art, daß der genaue Nenner von γ_n ein Primideal aus $\mathfrak{K}$ enthält, welches nicht in den Nennern der Zahlen $\gamma_0, \cdots, \gamma_{n-1}$ und $\gamma_0^{(\nu)}, \cdots, \gamma_{n+r}^{(\nu)} (\nu = 1, \cdots, h)$ aufgeht, so besteht zwischen $f, f_1, \cdots, f_h$ keine algebraische Gleichung, die f wirklich enthält.

Zum Beweise werde angenommen, daß zwischen $f, f_1, \cdots, f_h$ eine f enthaltende algebraische Gleichung $P = 0$ bestehe. Ihr Grad l in bezug auf f sei möglichst klein, so daß also $\frac{\partial P}{\partial f}$ nicht identisch in x verschwindet. Setzt man

$$P = p_0 f^l + \cdots + p_l,$$

so bestimmen sich die Koeffizienten der von $f_1, \cdots, f_h$ abhängigen Polynome $p_0, \cdots, p_l$ aus homogenen linearen Gleichungen, welche man erhält, indem man in der Relation $P = 0$

für $f, f_1, \cdots, f_h$ die Potenzreihen einträgt und Koeffizientenvergleichung ausführt. Daher können die Koeffizienten von $p_0, \cdots, p_l$ als ganze Zahlen von $\mathfrak{K}$ gewählt werden. Es sei nun

$$\frac{\partial P}{\partial f} = c_r x^r + \cdots$$

mit $c_r \neq 0$ die Potenzreihe für $\frac{\partial P}{\partial f}$. Setzt man noch

$$\sum_{n=0}^{r} \gamma_n x^n = g,$$

so ist nach dem TAYLORschen Satze

$$P(f, \cdots) = P(g, \cdots) + (f-g)\frac{\partial P(g, \cdots)}{\partial g} + \frac{(f-g)^2}{2}\frac{\partial^2 P(g, \cdots)}{\partial g^2} + \cdots = 0,$$

(19) $$(f-g)\frac{\partial P(g)}{\partial g} = -P(g) - \frac{(f-g)^2}{2}\frac{\partial^2 P(g)}{\partial g^2} + \cdots.$$

Da die Reihen f und g in den ersten $r+1$ Gliedern übereinstimmen, so ist auch

$$\frac{\partial P}{\partial g} = c_r x^r + \cdots.$$

In (19) vergleiche man nun die Koeffizienten von x^{n+r}, wo n eine natürliche Zahl $> r$ ist. Offenbar wird $c_r \gamma_n$ ein Polynom in $\gamma_0, \cdots, \gamma_{n-1}$ und $\gamma_0^{(\nu)}, \cdots, \gamma_{n+r}^{(\nu)}$ $(\nu = 1, \cdots, h)$ mit ganzen Koeffizienten aus $\mathfrak{K}$. Nach Voraussetzung läßt sich nun n so wählen, daß im genauen Nenner von γ_n ein Primideal $\mathfrak{p}_n$ aufgeht, welches nicht in den Nennern der Zahlen $\gamma_0, \cdots, \gamma_{n-1}$ und $\gamma_0^{(\nu)}, \cdots, \gamma_{n+r}^{(\nu)}$ enthalten ist. Dann enthält auch der Nenner von $c_r \gamma_n$ das Primideal $\mathfrak{p}_n$ nicht, und daher ist der Zähler von c_r durch $\mathfrak{p}_n$ teilbar. Da dies für unendliche viele n gilt und die zu verschiedenen n gehörigen Primideale $\mathfrak{p}_n$ ebenfalls verschieden sind, so entsteht ein Widerspruch.

Der zweite Beweis von Satz 3 verläuft nun folgendermaßen: Es sei $P(y_1, y_1', y_2, x) = 0$ eine algebraische Gleichung zwischen y_1, y_1', y_2, x, die y_2 wirklich enthält. Sind y_3, y_4 irgend zwei linear unabhängige Lösungen der BESSELschen Differentialgleichung, so ist mit konstanten p, q, r, s

$$y_1 = py_3 + qy_4, \quad y_2 = ry_3 + sy_4$$

und nach (14)

$$y_1' = py_3' + q\frac{y_4 y_3'}{y_3} + \frac{\alpha q}{ps - qr}\frac{1}{xy_3};$$

dadurch geht $P = 0$ in eine algebraische Gleichung zwischen y_3, y_3', y_4, x über, und umgekehrt läßt sich jede Gleichung zwischen y_3, y_3', y_4, x als Gleichung zwischen y_1, y_1', y_2, x schreiben. Nach Satz 2 genügt es also, unter y_1 und y_2 zwei spezielle nicht proportionale Lösungen von (7) zu verstehen.

Ist λ ganz rational, so gibt es eine bei $x = 0$ reguläre Lösung y_1 und eine bei $x = 0$ *logarithmisch* verzweigte Lösung y_2. Dem widerspricht, daß jede Lösung y der Gleichung $P(y_1, y_1', y, x) = 0$ bei $x = 0$ den Charakter einer *algebraischen* Funktion besitzt.

Man kann sich also auf den Fall beschränken, daß 2λ keine ganze Zahl ist. Setzt man

(20) $$K_\lambda(x) = \sum_{n=0}^{\infty} \frac{(-1)^n}{n!\,(\lambda+1)(\lambda+2)\cdots(\lambda+n)}\left(\frac{x}{2}\right)^{2n},$$

so ist

$$\frac{1}{\Gamma(\lambda+1)}\left(\frac{x}{2}\right)^{\lambda} K_\lambda(x) = J_\lambda(x)$$

die BESSELsche Funktion, und es sind J_λ und $J_{-\lambda}$ zwei linear unabhängige Lösungen von (7). Aus einer $J_{-\lambda}$ enthaltenden algebraischen Gleichung zwischen $J_\lambda, J_\lambda', J_{-\lambda}, x$ folgt eine algebraische Gleichung zwischen K_λ, K'_λ, $K_{-\lambda}$, x, x^λ, also, da K_λ und $K_{-\lambda}$ bei $x=0$ regulär sind, auch eine algebraische Gleichung

$$R(K_\lambda, K'_\lambda, K_{-\lambda}, x) = 0,$$

die wieder $K_{-\lambda}$ enthält. Die Koeffizienten $\xi_1, \cdots, \xi_q$ des Polynoms R bestimmen sich durch Koeffizientenvergleichung, also aus unendlich vielen homogenen linearen Gleichungen

$$\alpha_{k1}\xi_1 + \cdots + \alpha_{kq}\xi_q = 0, \qquad (k=0,1,\cdots) \tag{21}$$

in denen die Größen $\alpha_{kl} (k=0, 1, \cdots; l=1, \cdots, q)$ ganz rational mit ganzen rationalen Zahlenkoeffizienten aus den Koeffizienten von K_λ und $K_{-\lambda}$ gebildet und daher rationale Funktionen von λ mit rationalen Koeffizienten sind. Wären alle q-reihigen Determinanten der Matrix (α_{kl}) identisch in λ gleich 0, so wären $\xi_1, \cdots, \xi_q$ als Polynome in λ wählbar, und dann gälte die Gleichung $R=0$ ebenfalls identisch in λ. Gibt es eine nicht identisch verschwindende q-reihige Determinante, so ist diese eine rationale Funktion von λ mit rationalen Koeffizienten; sie muß 0 sein, da das System (21) lösbar ist, und demnach genügt λ einer algebraischen Gleichung mit rationalen Koeffizienten. In beiden Fällen kann man also annehmen, daß λ einem algebraischen Zahlkörper $\mathfrak{K}$ angehört, und dann sind auch die Koeffizienten von K_λ und $K_{-\lambda}$ Zahlen von $\mathfrak{K}$.

Zunächst sei λ rational, also $\lambda = \frac{a}{b}$, $(a, b) = 1$, $b \geq 3$. Nach dem Satz von Dirichlet gibt es unendlich viele Primzahlen p von der Form $bn - a$. Es sei $(b-1)n - a - 1 > 0$. Für $m = 1, 2, \cdots, (b-1)n - a - 1$ ist dann $0 < m+n < p$; also geht p nicht in $b(m+n) = p + bm + a$ auf, also auch nicht in $bm + a$, also nach (20) für $p > 2$ auch nicht in den Nennern der Koeffizienten von $x^0, x^1, \cdots, x^{2(b-1)n-2a-2}$ bei K_λ und K'_λ. Ferner ist p ein Faktor im Nenner des Koeffizienten von x^{2n} bei $K_{-\lambda}$, während die Nenner der vorhergehenden Koeffizienten zu p teilerfremd sind. Nach dem oben bewiesenen Hilfssatz ist dann aber $K_{-\lambda}$ nicht von K_λ, K'_λ, x algebraisch abhängig.

Nun sei λ algebraisch irrational, und es sei

$$f(x) = a_0(x-\lambda_1)\cdots(x-\lambda_s) = a_0 x^s + \cdots = 0$$

eine irreduzible Gleichung für λ mit ganzen rationalen Koeffizienten. Nach einem von Nagell bewiesenen Satze wird der größte Primteiler des Produktes $f(1)f(2)\cdots f(n)$ stärker unendlich als n selber. Insbesondere kann man also beliebig große natürliche Zahlen n finden, so daß $f(n)$ eine Primzahl p enthält, die größer als $3n$ ist und nicht in $f(1), \cdots, f(n-1)$ aufgeht. Außerdem sei noch $p > |a_0|$. Es sei $\mathfrak{p}$ ein Primideal aus $\mathfrak{K}$, welches in p und in $a_0(n-\lambda)$ aufgeht. Für $m = 1, 2, \cdots, 2n$ ist $0 < m+n \leq 3n$, also p kein Faktor von $m+n$, also auch nicht von $a_0(m+n) = a_0(m+\lambda) + a_0(n-\lambda)$. Das Primideal $\mathfrak{p}$ geht daher nicht in $a_0(m+\lambda)$ auf. Es ist also $\mathfrak{p}$ ein Faktor des Nenners des Koeffizienten von x^{2n} bei $K_{-\lambda}$, dagegen ist $\mathfrak{p}$ nicht in den Nennern der Koeffizienten von $x^0, \cdots, x^{2n-1}$ bei $K_{-\lambda}$ und von $x^0, \cdots, x^{4n}$ bei K_λ und K'_λ enthalten. Nach dem Hilfssatz ist folglich $K_{-\lambda}$ auch in diesem Falle von K_λ, K'_λ, x algebraisch unabhängig.

Damit ist Satz 3 ein zweites Mal bewiesen. Der Spezialfall der algebraischen Unabhängigkeit von J_λ, $J_{-\lambda}$, x hat sich dabei sogar ohne Benutzung von Satz 2 ergeben.

Satz 3 ermöglicht nun die Anwendung des ersten Hilfssatzes der Einleitung. Es seien ν verschiedene ganze nicht negative Zahlen $l_1, \cdots, l_\nu$ gegeben und ferner $(l_1+1) + \cdots + (l_\nu+1) = q$ Polynome $f_{kl}(x)$, wo die Indizes die Werte $k = 0, \cdots, l$ und $l = l_1, \cdots, l_\nu$ besitzen. Für kein l der Reihe $l_1, \cdots, l_\nu$ seien die $l+1$ Polynome $f_{0l}, \cdots, f_{ll}$ sämtlich identisch 0. Mit einer Lösung y der Besselschen Differentialgleichung bilde man nun für $l = l_\varrho$ und $\varrho = 1, \cdots, \nu$ den Ausdruck

$$\phi_\varrho = f_{0l}y'^l + f_{1l}y'^{l-1}y + \cdots + f_{ll}y^l \tag{22}$$

und die Summe

$$\phi = \phi_1 + \phi_2 + \cdots + \phi_\nu = P(y, y'). \tag{23}$$

Dies ist ein Polynom in y und y', das nur die Dimensionen $l_1, \cdots, l_\nu$ wirklich enthält; seine Koeffizienten sind die q Polynome f_{kl}. Die q Funktionen $w_{kl} = y^k y'^{l-k}$ $(k=0,$

$\cdots, l;\ l = l_1, \cdots, l_\nu)$ genügen einem System von q homogenen linearen Differentialgleichungen erster Ordnung; es ist nämlich auf Grund von (7)

$$(24)\qquad \frac{dw_{kl}}{dx} = kw_{k-1,l} - \frac{l-k}{x} w_{kl} - (l-k)\left(1 - \frac{\lambda^2}{x^2}\right) w_{k+1,l}.$$

Da in jeder Differentialgleichung immer nur Funktionen mit demselben Index l auftreten, so zerfällt das System in ν einzelne voneinander ganz unabhängige Systeme, und in einem solchen Teilsystem treten nur die Funktionen $w_{0l}, \cdots, w_{ll}$ auf. Die Gleichungen (24) sind nun aber erfüllt, wenn in $w_{kl} = y^k y'^{l-k}$ für y irgendeine Lösung der Besselschen Differentialgleichung gesetzt wird, also $y = \lambda_1 y_1 + \lambda_2 y_2$, wo y_1 und y_2 zwei linear unabhängige Lösungen und λ_1, λ_2 beliebige Konstanten bedeuten. Setzt man zur Abkürzung noch

$$(25)\qquad \sum_{\rho=0}^{l} \binom{k}{\rho}\binom{l-k}{r-\rho} y_1^\rho y_2^{k-\rho} y_1'^{r-\rho} y_2'^{l-k-r+\rho} = \psi_{krl}, \qquad (r = 0, \cdots, l)$$

so ist

$$(26)\qquad w_{kl} = (\lambda_1 y_1 + \lambda_2 y_2)^k (\lambda_1 y_1' + \lambda_2 y_2')^{l-k} = \sum_{r=0}^{l} \lambda_1^r \lambda_2^{l-r} \psi_{krl}, \qquad (k = 0, \cdots, l)$$

und da diese Funktionen identisch in λ_1, λ_2 den Gleichungen (24) genügen, so ist auch für jedes feste l

$$w_{kl} = \sum_{r=0}^{l} c_r \psi_{krl} \qquad (k = 0, \cdots, l)$$

eine Lösung von (24), wo $c_0, \cdots c_l$ irgendwelche Konstanten bedeuten. Dies ist aber die allgemeine Lösung, denn aus

$$\sum_{r=0}^{l} c_r \psi_{krl} = 0 \qquad (k = 0, \cdots, l)$$

folgt speziell für $k = l$ die Gleichung

$$\sum_{r=0}^{l} c_r \binom{k}{r} y_1^r y_2^{k-r} = 0,$$

also wegen der linearen Unabhängigkeit von y_1, y_2

$$c_0 = 0, \cdots, c_l = 0.$$

Damit ist auch die allgemeine Lösung des vollen Systems (24) gefunden.

Nach (22), (23) und (24) sind die Funktionen $\frac{d^s \phi}{dx^s} = \phi^{(s)}$ für $s = 1, 2, \cdots$ wieder Polynome in y, y', die nur Glieder der Dimensionen $l_1, \cdots, l_\nu$ wirklich enthalten, also homogene lineare Formen der w_{kl}. Und nun soll bewiesen werden, daß die Determinante der q für $s = 0, 1, \cdots, q-1$ entstehenden Formen $\phi, \phi', \cdots, \phi^{(q-1)}$ nicht identisch verschwindet. Nach dem ersten Hilfssatz der Einleitung hat man nur zu zeigen, daß die q Funktionen, welche aus

$$\phi = \sum_{k,l} f_{kl} w_{kl}$$

dadurch entstehen, daß man für $l = l_\rho$ die Funktionen $w_{0l}, \cdots, w_{ll}$ durch $\psi_{orl}, \cdots, \psi_{lrl}$, für $l \neq l_\rho$ durch 0 ersetzt und $\rho = 1, \cdots, \nu$ sowie $r = 0, \cdots, l$ wählt, linear unabhängig sind.

Gilt eine Gleichung

$$(27)\qquad \sum_{l} \sum_{r=0}^{l} \sum_{k=0}^{l} c_{rl} f_{kl} \psi_{krl} = 0$$

mit konstanten c_{rl}, so erhält man hieraus eine algebraische Gleichung zwischen y_1, y_1', y_2, x, indem man vermöge (14) die Funktion y_2' eliminiert. Nach Satz 3 bekommt man dann eine Identität in den 4 Variabeln y_1, y_1', y_2, x. In dieser sollen nur die Glieder höchster

Dimension in y_1, y_1', y_2 betrachtet werden. Nach (14) erhält man diese, indem man in (27) nur das größte l beibehält, für welches die Konstanten $c_{ol}, \cdots, c_{ll}$ nicht sämtlich o sind, und in dem durch (25) gegebenen Ausdruck von ψ_{krl} die Funktion y_2' durch $\frac{y_2}{y_1} y'$ ersetzt. Beachtet man (26), so folgt

$$\sum_{r=0}^{l} \sum_{k=0}^{l} c_{rl} f_{kl} \binom{l}{r} y_1^r y_2^{l-r} \left(\frac{y_1'}{y_1}\right)^{l-k} = 0,$$

identisch in $y_1, y_2, \frac{y_1'}{y_1}$. Daher ist

$$c_{rl} f_{kl} = 0, \qquad (r = 0, \cdots, l;\ k = 0, \cdots, l)$$

und dies ist ein Widerspruch, da weder die Konstanten $c_{ol}, \cdots, c_{ll}$ noch die Polynome $f_{ol}, \cdots, f_{ll}$ sämtlich identisch o sind. Folglich gilt

Satz 4:

Es sei λ nicht die Hälfte einer ungeraden Zahl und y eine Lösung der Besselschen *Differentialgleichung* (7). *Man bilde den Ausdruck*

$$\phi = \sum_l \sum_{k=0}^{l} f_{kl} y^k y'^{l-k},$$

dessen Koeffizienten f_{kl} Polynome in x seien und in dem nur die Dimensionen $l = l_1, \cdots, l_\nu$ in y, y' wirklich auftreten mögen, so daß also ϕ eine homogene lineare Form der $(l_1 + 1) + \cdots + (l_\nu + 1) = q$ Potenzprodukte $y^k y'^{l-k}$ $(k = 0, \cdots, l;\ l = l_1, \cdots, l_\nu)$ ist. Dann ist auch jede Ableitung von ϕ eine solche homogene lineare Form, und die Determinante der q Formen $\phi, \phi', \cdots, \phi^{(q-1)}$ verschwindet nicht identisch.

Offenbar läßt sich dieser Satz auch auf die Lösungen anderer homogener linearer Differentialgleichungen zweiter Ordnung mit algebraischen Koeffizienten übertragen.

§ 2. Arithmetische Hilfsmittel.

Um die Anwendung des zweiten Hilfssatzes der Einleitung zu ermöglichen, sind jetzt die Koeffizienten der Funktionen $y^k y'^{l-k}$ arithmetisch zu untersuchen. Das Mittel hierzu wird geliefert durch den von Maier stammenden

Satz 5:

Es seien α und γ rationale Zahlen; γ sei verschieden von $0, -1, -2, \cdots$. Es sei h_n der Hauptnenner der n Brüche

$$\frac{\alpha}{\gamma}, \frac{\alpha(\alpha+1)}{\gamma(\gamma+1)}, \cdots, \frac{\alpha(\alpha+1)\cdots(\alpha+n-1)}{\gamma(\gamma+1)\cdots(\gamma+n-1)}. \qquad (n = 1, 2, \cdots)$$

Dann wächst h_n schwächer an als die n-te Potenz einer geeigneten Konstanten.

Beweis: Es sei $\alpha = a : b$, $\gamma = c : d$, $(c, d) = 1$, $d > 0$. Wegen der Gleichung

$$\text{(28)} \qquad \frac{\alpha(\alpha+1)\cdots(\alpha+l-1)}{\gamma(\gamma+1)\cdots(\gamma+l-1)} \cdot \frac{b^{2l}}{d^l} = \frac{a(a+b)\cdots(a+(l-1)b)b^l}{c(c+d)\cdots(c+(l-1)d)} \qquad (l = 1, 2, \cdots, n)$$

genügt es, den Satz für den Hauptnenner der rechten Seite von (28) zu beweisen.

Der Nenner $c(c+d)\cdots(c+(l-1)d) = N_l$ ist zu d teilerfremd. Es sei p ein Primfaktor von N_l. Durchläuft ν irgend p^k konsekutive ganze rationale Zahlen, so ist von den p^k Zahlen $c + \nu d$ genau eine durch p^k teilbar. Von den l Faktoren des Nenners N_l sind daher mindestens $[lp^{-k}]$ und höchstens $[lp^{-k}] + 1$ durch p^k teilbar; für $p^k > |c| + (l-1)d$ ist aber keiner durch p^k teilbar. Für den Exponenten s der in N_l aufgehenden Potenz von p gilt daher die Ungleichung

$$\sum_k [lp^{-k}] \leq s \leq \sum_k ([lp^{-k}] + 1),$$

und hierin durchläuft k alle natürlichen Zahlen, welche der Bedingung $p^k \leq |c| + (l-1)d$ genügen. Daher ist mit konstanten c_1 und c_2

$$\left[\frac{l}{p}\right] \leq s < \left[\frac{l}{p}\right] + c_1 \frac{l}{p^2} + c_2 \frac{\log l}{\log p}.$$

Im Zähler $a(a+b)\cdots(a+(l-1)b)b^l = Z_l$ geht p mindestens zur $\left[\frac{l}{p}\right]$-ten Potenz auf. Folglich ist im Nenner des reduzierten Bruches $Z_l : N_l$ der Exponent von p kleiner als $c_1 \frac{l}{p^2} + c_2 \frac{\log l}{\log p}$. Man erhält für den Logarithmus des Hauptnenners H_n der n Brüche $Z_1 : N_1, \cdots, Z_n : N_n$ die Abschätzung

$$\log H_n < \sum_p \left(c_1 \frac{n}{p^2} + c_2 \frac{\log n}{\log p}\right) \log p = c_1 n \sum_p \frac{\log p}{p^2} + c_2 \log n \sum_p 1,$$

wo p alle Primzahlen unterhalb $|c| + (n-1)d$ durchläuft. Nach einem elementar beweisbaren Satz der Primzahllehre ist nun

$$\sum_p 1 < c_3 \frac{n}{\log n};$$

ferner konvergiert die über alle Primzahlen erstreckte Summe $\sum \frac{\log p}{p^2} = c_4$. Daher ist

$$\log H_n < c_1 c_4 n + c_2 c_3 n = (c_1 c_4 + c_2 c_3) n,$$

was den Satz beweist.

Nunmehr betrachte man Potenzreihen

$$y = \sum_{n=0}^{\infty} \frac{a_n}{b_n} \frac{x^n}{n!}$$

mit folgenden Eigenschaften:

1. Die Zähler $a_0, a_1, \cdots$ sind ganze Zahlen eines festen algebraischen Zahlkörpers, und die absoluten Beträge sämtlicher Konjugierten von a_n wachsen mit n schwächer an als jede feste positive Potenz von $n!$;

2. Die Nenner $b_0, b_1, \cdots$ sind natürliche Zahlen, und das kleinste gemeinschaftliche Vielfache von $b_0, \cdots, b_n$ wächst mit n ebenfalls schwächer an als jede feste positive Potenz von $n!$;

3. Die Funktion y genügt einer linearen Differentialgleichung, deren Koeffizienten Polynome mit algebraischen Zahlenkoeffizienten sind.

Eine Funktion y, deren Potenzreihe diese drei Eigenschaften hat, möge kurz eine E-Funktion genannt werden. Offenbar ist die Exponentialfunktion eine E-Funktion. Jede E-Funktion ist ganz. Die E-Funktionen haben einige wichtige, zum Teil evidente Eigenschaften, welche nun angeführt werden sollen; dabei bedeute $E(x)$ eine beliebige E-Funktion.

I. Jede algebraische Konstante ist eine E-Funktion.
II. Für algebraisches konstantes α ist $E(\alpha x)$ eine E-Funktion.
III. Die Ableitung $E'(x)$ ist eine E-Funktion.
IV. Das Integral $\int_0^x E(t)\,dt$ ist eine E-Funktion.
V. Mit $E_1(x)$ und $E_2(x)$ ist auch $E_1(x) + E_2(x)$ eine E-Funktion.
VI. Mit $E_1(x)$ und $E_2(x)$ ist auch $E_1(x)\,E_2(x)$ eine E-Funktion.

Von diesen Behauptungen bedürfen nur die unter V. und VI. eines Beweises. Es sei

(29) $$E_1(x) = \sum_{n=0}^{\infty} \frac{a_n'}{b_n'} \frac{x^n}{n!}, \quad E_2(x) = \sum_{n=0}^{\infty} \frac{a_n''}{b_n''} \frac{x^n}{n!},$$

also, wenn

$$(30)\qquad a_n' b_n'' + a_n'' b_n' = a_n, \quad b_n' b_n'' = b_n, \quad E_1(x) + E_2(x) = y$$

gesetzt wird,

$$y = \sum_{n=0}^{\infty} \frac{a_n}{b_n} \frac{x^n}{n!}.$$

Da a_n', a_n'', b_n', b_n'' die Bedingungen 1. und 2. erfüllen, so erfüllt auch a_n die Bedingung 1. Bedeutet $\{p, \cdots, q\}$ für natürliche Zahlen $p, \cdots, q$ ihr kleinstes gemeinschaftliches Vielfache, so ist

$$\{b_0' b_0'', \cdots, b_n' b_n''\} \leq \{b_0', \cdots, b_n'\}\{b_0'', \cdots, b_n''\},$$

also erfüllt b_n die Bedingung 2. Sind ferner die linearen Differentialgleichungen für E_1 und E_2 von den Ordnungen h_1 und h_2, so ist jede Ableitung von y eine lineare Kombination von $E_1, E_1', \cdots, E_1^{(h_1-1)}, E_2, E_2', \cdots, E_2^{(h_2-1)}$, und zwar treten als Koeffizienten rationale Funktionen von x auf, die mit algebraischen Zahlenkoeffizienten versehen sind. Folglich genügt y einer linearen Differentialgleichung von der Ordnung $h_1 + h_2$, deren Koeffizienten Polynome mit algebraischen Zahlenkoeffizienten sind. Damit ist V. bewiesen.

Zum Beweise von VI. setze man abweichend von (30)

$$(31)\qquad \{b_0', \cdots, b_n'\}\{b_0'', \cdots, b_n''\} = b_n$$

$$(32)\qquad b_n \sum_{k=0}^{n} \binom{n}{k} \frac{a_k' a_{n-k}''}{b_k' b_{n-k}''} = a_n,$$

dann ist mit (29)

$$y = E_1 E_2 = \sum_{m=0}^{\infty} \sum_{n=0}^{\infty} \binom{m+n}{n} \frac{a_m' a_n''}{b_m' b_n''} \frac{x^{m+n}}{(m+n)!} = \sum_{n=0}^{\infty} \frac{a_n}{b_n} \frac{x^n}{n!}.$$

Es ist $\{b_0, \cdots, b_n\} = b_n$ und nach (31) die Bedingung 2. erfüllt. Mit Rücksicht auf $\sum_{k=0}^{n} \binom{n}{k} = 2^n$ und (32) ist auch Bedingung 1. erfüllt. Endlich ist Bedingung 3. erfüllt, weil sich jede Ableitung von y linear durch die $h_1 h_2 + h_1 + h_2$ Funktionen $E_1, E_1', \cdots, E_1^{(h_1-1)}, E_2, E_2', \cdots, E_2^{(h_2-1)}, E_1 E_2, \cdots, E_1^{(h_1-1)} E_2^{(h_2-1)}$ ausdrücken läßt.

Aus I., $\cdots$, VI. folgt nun

Satz 6:

Es seien $E_1(x), \cdots, E_m(x)$ irgendwelche E-Funktionen und $\alpha_1, \cdots, \alpha_m$ algebraische Zahlen. Dann ist jedes mit algebraischen Koeffizienten gebildete Polynom in $E_1(\alpha_1 x), \cdots, E_m(\alpha_m x)$ und den Ableitungen dieser Funktionen eine E-Funktion.

Jedes Polynom in x mit algebraischen Koeffizienten ist insbesondere eine E-Funktion. Ein nichttriviales Beispiel von E-Funktionen erhält man durch Verwendung von Satz 5. Man nehme l rationale Zahlen $\gamma_1, \cdots, \gamma_l$, die sämtlich von $0, -1, -2, \cdots$ verschieden sind, und k rationale Zahlen $\alpha_1, \cdots, \alpha_k$. Es sei $l - k = t > 0$. Man setze $c_n = 0$, falls n kein Multiplum von t ist, aber

$$c_n = \prod_{p=1}^{k} \left(\alpha_p(\alpha_p+1) \cdots (\alpha_p+m-1)\right) : \prod_{q=1}^{l} \left(\gamma_q(\gamma_q+1) \cdots (\gamma_q+m-1)\right),$$

falls $n = mt$ durch t teilbar ist. Um einzusehen, daß

$$(33)\qquad y = \sum_{n=0}^{\infty} c_n x^n$$

eine E-Funktion ist, schreibe man c_n für $n = mt$ in der Gestalt

$$(34)\qquad c_n = \frac{\alpha_1 \cdots (\alpha_1+m-1)}{\gamma_1 \cdots (\gamma_1+m-1)} \cdots \frac{\alpha_k \cdots (\alpha_k+m-1)}{\gamma_k \cdots (\gamma_k+m-1)} \cdot \frac{1 \cdots m}{\gamma_{k+1} \cdots (\gamma_{k+1}+m-1)} \cdots \frac{1 \cdots m}{\gamma_l \cdots (\gamma_l+m-1)} \cdot \frac{(mt)!}{(m!)^t} \cdot \frac{1}{n!}$$

Bedeutet nun $\frac{a_n}{b_n}$ den reduzierten Bruch $n!\, c_n$, so erfüllen a_n und b_n nach Satz 5 die Bedingungen 1. und 2. Daß auch die Bedingung 3. erfüllt ist, ergibt sich aus der Form der c_n, welche leicht die lineare Differentialgleichung für y abzuleiten gestattet.

Also ist die Besselsche Funktion

$$J_0(x) = \sum_{n=0}^{\infty} \frac{(-1)^n}{n!\, n!} \left(\frac{x}{2}\right)^{2n}$$

eine E-Funktion und allgemeiner ist auch die Funktion

$$K_\lambda(x) = \Gamma(\lambda+1) \left(\frac{x}{2}\right)^{-\lambda} J_\lambda(x) = \sum_{n=0}^{\infty} \frac{(-1)^n}{n!(\lambda+1)\cdots(\lambda+n)} \left(\frac{x}{2}\right)^{2n}$$

für jedes von $-1, -2, \cdots$ verschiedene rationale λ eine E-Funktion.

Es wäre nicht ohne Interesse, E-Funktionen anzugeben, welche nicht aus den speziellen, durch (33) und (34) definierten E-Funktionen durch die in I., $\cdots$, VI. angegebenen Operationen gewonnen werden können.

Man kann nun nach algebraischen Relationen zwischen gegebenen E-Funktionen $E_1(x), \cdots, E_m(x)$ fragen, also nach Polynomen in den Variabeln $x_1, \cdots, x_m$, die identisch in x verschwinden, wenn für $x_1, \cdots, x_m$ die Funktionen $E_1(x), \cdots, E_m(x)$ eingesetzt werden. Eine solche Relation ist z. B.

$$x^2 K^2_{\frac{1}{2}} + K^2_{-\frac{1}{2}} = 1 .$$

Durch Satz 3 ist ein ganz spezieller Fall dieser allgemeinen Problemstellung erledigt worden, indem dort gezeigt wurde, daß zwischen $K_\lambda(x)$, $K'_\lambda(x)$, $K_{-\lambda}(x)$, x für $\lambda \neq 0, \pm 1, \pm 2, \cdots$ dann und nur dann eine algebraische Gleichung gilt, wenn 2λ ungerade ist. Ein etwas weitergehender Satz über die Besselschen Funktionen wird sich später ergeben.

Viel tiefer scheint das folgende Problem gelegen zu sein: Es seien $\alpha_1, \cdots, \alpha_m$ algebraische Zahlen; es soll festgestellt werden, ob zwischen den *Zahlen* $E_1(\alpha_1), \cdots, E_m(\alpha_m)$ eine algebraische Gleichung mit rationalen Koeffizienten besteht. Dieses Problem enthält das obengenannte, denn die identisch in x erfüllten Gleichungen zwischen $E_1(x), \cdots, E_m(x)$ lassen sich mit algebraischen Koeffizienten schreiben und sind insbesondere für algebraisches x erfüllt. Man kann es noch etwas anders formulieren. Jedes Potenzprodukt von $E_1(\alpha_1 x), \cdots, E_m(\alpha_m x)$ ist nämlich wieder eine E-Funktion; das Problem geht also über in die Aufgabe, zu entscheiden, ob die Werte $E_1(1), \cdots, E_m(1)$ im Körper der rationalen Zahlen linear abhängig sind. Die Behandlung dieser Aufgabe erfolgt nach der in der Einleitung skizzierten Methode, deren Durchführbarkeit nur das Bestehen einer Aussage vom Typus des Satzes 4 erfordert. Damit geht dann das arithmetische Problem der Zahlengleichung über in das algebraische Problem der identisch in x bestehenden Funktionalgleichung. Als Beispiel für diese Bemerkung diene die Exponentialfunktion; nach dem Lindemannschen Satz sind alle algebraischen Gleichungen $P(e^{\alpha_1}, \cdots, e^{\alpha_m}) = 0$ algebraische Folgen der Funktionalgleichung $\exp(x+y) = \exp x \exp y$.

Es sollen nun Näherungsformen für E-Funktionen angegeben werden, die sowohl algebraisch als auch arithmetisch eine gute Approximation von 0 liefern.

Satz 7:

Es seien k E-Funktionen $E_1(x), \cdots, E_k(x)$ mit rationalen Koeffizienten gegeben. Es sei n eine natürliche Zahl. Es gibt k Polynome $P_1(x), \cdots, P_k(x)$ vom Grade $2n-1$ mit folgenden Eigenschaften:

1. *Die Koeffizienten von $P_1(x), \cdots, P_k(x)$ sind ganz rational, nicht sämtlich 0 und als Funktion von n höchstens von der Größenordnung $(n!)^{2+\varepsilon}$, wo ε eine beliebig kleine feste positive Zahl bedeutet;*

2. *es ist*

$$P_1 E_1 + \cdots + P_k E_k = \sum_{\nu=(2k-1)n}^{\infty} q_\nu \frac{x^\nu}{\nu!}, \tag{35}$$

so daß also die links stehende E-Funktion $P_1E_1 + \cdots + P_kE_k$ bei $x = 0$ mindestens von der Ordnung $(2k-1)n$ verschwindet;

3. *die Koeffizienten q_ν sind als Funktion von n und ν höchstens von der Größenordnung $(n!)^2(\nu!)^\varepsilon$.*

Beweis: Es sei

$$E(x) = \sum_{n=0}^{\infty} \gamma_n \frac{x^n}{n!}$$

eine der k Funktionen $E_1, \cdots, E_k$. Man setze mit ganzen rationalen $g_0, \cdots, g_{2n-1}$

(36) $$P(x) = (2n-1)! \sum_{\nu=0}^{2n-1} g_\nu \frac{x^\nu}{\nu!},$$

(37) $$d_l = \sum_{\rho=0}^{2n-1} \binom{l}{\rho} g_\rho \gamma_{l-\rho};$$

dann ist

(38) $$P(x)E(x) = (2n-1)! \sum_{l=0}^{\infty} d_l \frac{x^l}{l!}.$$

Sollen nun in der Potenzreihe für $P_1E_1 + \cdots + P_kE_k$ die Koeffizienten von $x^0, x^1, \cdots, x^{(2k-1)n-1}$ sämtlich 0 sein, so müssen $(2k-1)n$ homogene lineare Gleichungen für die $2kn$ unbekannten Koeffizienten der k Polynome $P_1, \cdots, P_k$ vom Grade $2n-1$ erfüllt sein. Der Hauptnenner der rationalen Zahlen $\gamma_0, \cdots, \gamma_{(2k-1)n-1}$ ist $O((n!)^\varepsilon)$; dieselbe Abschätzung gilt für die Binomialkoeffizienten $\binom{l}{\rho}$ mit $\rho = 0, \cdots, l$ und $l = 0, \cdots, (2k-1)n-1$ und daher auch für die ganzen rationalen Koeffizienten der $(2k-1)n$ homogenen linearen Gleichungen. Nach dem zweiten Hilfssatz der Einleitung sind diese Gleichungen lösbar in solchen ganzen rationalen Werten der unbekannten g_ν, welche nicht sämtlich 0 und von der Größenordnung

(39) $$\left(2kn(n!)^\varepsilon\right)^{\frac{(2k-1)n}{2kn-(2k-1)n}}$$

sind.

Die so bestimmten Polynome $P_1, \cdots, P_k$ haben die behaupteten drei Eigenschaften. Nach (36) und (39) ist nämlich 1. erfüllt; ferner ist 2. erfüllt; und da $\gamma_\nu = O((\nu!)^\varepsilon)$ ist, so folgt aus (37) und (38) für den Koeffizienten q_ν von $\frac{x^\nu}{\nu!}$ auf der rechten Seite von (35) die in 3. ausgesprochene Abschätzung.

Die Bedeutung von Satz 7 liegt darin, daß einerseits die Entwicklung (35) mit einer *hohen* Potenz von x beginnt und daß andererseits die ganzen rationalen Koeffizienten der Polynome $P_1, \cdots, P_k$ *klein* sind. Man könnte zwar noch erreichen, daß die Entwicklung (35) erst mit der Potenz x^{2kn-1} beginnt, dann würden aber vielleicht die Koeffizienten von $P_1, \cdots, P_k$ nicht mehr so klein sein, wie bei Satz 7 unter 1. ausgesagt wird; und gerade dies ist für das Folgende wesentlich.

Es sei l eine natürliche Zahl und $k = \frac{(l+1)(l+2)}{2}$. Unter $E_1(x), \cdots, E_k(x)$ verstehe man nunmehr die k Funktionen

$$J_0^\varkappa J_0'^{\lambda-\varkappa} \qquad (\varkappa = 0, \cdots, \lambda;\ \lambda = 0, \cdots, l)$$

und wende auf sie den Satz 7 an. Für jedes n gibt es also k nicht sämtlich identisch verschwindende Polynome $f_{\varkappa\lambda}$ $(\varkappa = 0, \cdots, \lambda;\ \lambda = 0, \cdots, l)$ vom Grade $2n-1$ mit ganzen rationalen Koeffizienten von der Größenordnung $(n!)^{2+\varepsilon}$, so daß die Potenzreihe für die Funktion

(40) $$\varphi(x) = \sum_{\varkappa,\lambda} f_{\varkappa\lambda}(x) J_0^\varkappa J_0'^{\lambda-\varkappa}$$

mit der Potenz $x^{(2k-1)n}$ beginnt und als Majorante

(41) $$O\left((n!)^2 \sum_{\nu=(2k-1)n} \frac{|x|^\nu}{(\nu!)^{1-\varepsilon}}\right)$$

besitzt.

Wegen der Differentialgleichung $xJ_0'' = -J_0' - xJ_0$ sind die Funktionen $x^a\phi^{(a)}(x)$ für $a = 1, 2, \cdots$ Polynome in x, J_0, J_0', und zwar in x vom Grade $2n+a-1$, in J_0, J_0' von der Dimension l. Die Koeffizienten dieser Polynome sind wieder ganz rational und für $a < n+k^2$ von der Größenordnung $(n!)^{1+\varepsilon}\cdot(n!)^{2+\varepsilon}$. Die Potenzreihe für $x^a\phi^{(a)}(x)$ beginnt ebenfalls mit $x^{(2k-1)n}$, und als Majorante erhält man die mit $n!$ multiplizierte Majorante (41). Die k Potenzprodukte $J_0^\varkappa J_0'^{\lambda-\varkappa}$ mögen in irgendeiner Reihenfolge mit $t_1, \cdots, t_k$ bezeichnet werden; dann ist $x^a\phi^{(a)}(x)$ eine homogene lineare Funktion von $t_1, \cdots, t_k$, etwa

$$x^a\phi^{(a)}(x) = \sigma_{a1}(x)t_1 + \cdots + \sigma_{ak}(x)t_k.$$

Fortan sei noch

(42) $$2n \geq k^2.$$

Satz 8:

Es sei ξ eine von 0 verschiedene Zahl. Unter den $n+k^2$ linearen Formen

$$\sigma_{a1}(\xi)t_1 + \cdots + \sigma_{ak}(\xi)t_k \qquad (a = 0, 1, \cdots, n+k^2-1)$$

der k Variabeln $t_1, \cdots, t_k$ gibt es k voneinander linear unabhängige Formen.

Beweis: Man hat Satz 4 anzuwenden und die Überlegungen vom Schlusse der Einleitung zu wiederholen. In der durch (40) definierten Funktion $\phi(x)$ mögen die Variabeln J_0 und J_0' nur in den Dimensionen $l_1, l_2, \cdots, l_\nu$ auftreten. Es kommen dann also nur die $(l_1+1)+\cdots+(l_\nu+1) = q$ Potenzprodukte $J_0^\varkappa J_0'^{\lambda-\varkappa}$ $(\varkappa = 0, \cdots, \lambda;\ \lambda = l_1, l_2, \cdots, l_\nu)$ vor, welche mit $\omega_1, \cdots, \omega_q$ bezeichnet seien. Es wird

(43) $$x^a\phi^{(a)}(x) = \tau_{a1}(x)\omega_1 + \cdots + \tau_{aq}(x)\omega_q,$$

wo $\tau_{a1}, \cdots, \tau_{aq}$ Polynome in x vom Grade $2n+a-1$ sind. Die Determinante

$$D(x) = |\tau_{ab}(x)|,$$

wo der Zeilenindex die Werte $a = 0, \cdots, q-1$, der Spaltenindex die Werte $b = 1, \cdots, q$ durchläuft, hat in x den Grad

$$q(2n-1)+1+2+\cdots+(q-1).$$

Bedeutet $T_{ba}(x)$ die Unterdeterminante von $\tau_{ab}(x)$, so ist nach (43) für $b = 1, \cdots, q$

(44) $$D(x)\omega_b = T_{b0}(x)\phi(x) + T_{b1}(x)x\phi'(x) + \cdots + T_{b\,q-1}(x)x^{q-1}\phi^{(q-1)}(x).$$

Nach Satz 4 ist die Determinante $D(x)$ nicht identisch gleich 0. Ferner verschwindet $x^a\phi^{(a)}(x)$ für $x = 0$ mindestens von der Ordnung $(2k-1)n$; dies tut also auch die rechte Seite von (44). Ist b so gewählt, daß ω_b eine der Potenzen $J_0^{l_1}, \cdots, J_0^{l_\nu}$ ist, so verschwindet ω_b nicht für $x = 0$. Daher ist $D(x)$ durch $x^{(2k-1)n}$ teilbar, und es gilt

(45) $$q(2n-1)+1+2+\cdots+(q-1)-(2k-1)n = \delta \geq 0.$$

Für $x = \xi \neq 0$ verschwinde $D(x)$ von der Ordnung s; dann ist $s \leq \delta$ und $D(\xi) = 0, \cdots, D^{(s-1)}(\xi) = 0$, $D^{(s)}(\xi) \neq 0$.

Nun ist ferner

(46) $$q = (l_1+1)+\cdots+(l_\nu+1) \leq 1+\cdots+(l+1) = k,$$

und hier steht das Gleichheitszeichen nur dann, wenn die Zahlen $l_1, \cdots, l_\nu$ mit den Zahlen $0, \cdots, l$ übereinstimmen, wenn also $\phi(x)$ alle Dimensionen in J_0, J_0' von der 0-ten bis zur l-ten wirklich enthält. Nach (45) ist

$$q \geq k - \tfrac{1}{2} + \frac{k - \frac{1}{2} - (1+\cdots+q-1)}{2n-1},$$

also nach (42) und (46)

$$q \geq k - \tfrac{1}{2} - \frac{k^2-3k+1}{2(2n-1)} > k-1,$$

also nach (46)

(47) $$q = k.$$

Daher treten in $\phi(x)$ wirklich alle Dimensionen $0, \cdots, l$ in J_0, J_0' auf; man kann die Funktionen $\omega_1, \cdots, \omega_q$ mit $t_1, \cdots, t_k$ und die Polynome $\tau_{a1}, \cdots, \tau_{aq}$ mit den Polynomen $\sigma_{a1}, \cdots, \sigma_{ak}$ identifizieren.

Wie in der Einleitung schließt man nun aus der s-ten Ableitung der Gleichung (44), daß von den $k+s$ linearen Formen

$$\sigma_{a1}(\xi)t_1 + \cdots + \sigma_{ak}(\xi)t_k \qquad (a = 0, \cdots, k+s-1)$$

der k Variabeln $t_1, \cdots, t_k$ sicherlich k voneinander unabhängig sind. Aus (45) und (47) ergibt sich noch

$$k+s \leq k+\delta = k(2n-1)+1+2+\cdots+k-(2k-1)n < n+k^2,$$

womit alles bewiesen ist.

§ 3. Die Transzendenz von $J_0(\xi)$.

Es sei ξ eine von 0 verschiedene Zahl. Man wähle, was nach Satz 8 möglich ist, k Zahlen $h_1, \cdots, h_k$ aus der Reihe $0, 1, \cdots, n+k^2-1$ so aus, daß die k linearen Formen

$$\phi_\nu = \sigma_{h_\nu 1}(\xi)t_1 + \cdots + \sigma_{h_\nu k}(\xi)t_k \qquad (\nu = 1, \cdots, k)$$

linear unabhängig sind.

Es sei $g(y, z)$ ein Polynom der Variabeln y, z von der Dimension $p \leq l$, dessen Koeffizienten ganz rational und nicht sämtlich 0 sind. Es sei G eine obere Schranke für den absoluten Betrag aller Koeffizienten von $g(y, z)$. Man setze $l-p = r$ und bilde die $\frac{(r+1)(r+2)}{2} = v$ Polynome $y^\varrho z^{\sigma-\varrho} g(y, z)$ mit $\varrho = 0, \cdots, \sigma$ und $\sigma = 0, \cdots, r$, deren Dimensionen $p+\sigma$ sämtlich $\leq l$ sind. Für die speziellen Werte $y = J_0(x)$, $z = J_0'(x)$ sind diese Polynome homogen linear in den Potenzprodukten $J_0^\varkappa J_0'^{\lambda-\varkappa}$ ($\varkappa = 0, \cdots, \lambda$; $\lambda = 0, \cdots, l$), also in $t_1, \cdots, t_k$; es entstehen so v neue lineare Formen von $t_1, \cdots, t_k$, etwa $\psi_1, \cdots, \psi_v$, deren Koeffizienten ganz rational und absolut $\leq G$ sind. Die v Polynome $y^\varrho z^{\sigma-\varrho} g(y, z)$ sind linear unabhängig voneinander, also sind dies auch $\psi_1, \cdots, \psi_v$. Aus den k linear unabhängigen Formen ϕ_ν ($\nu = 1, \cdots, k$) wähle man nun $k-v = w$ geeignete Formen, etwa ϕ_ν für $\nu = 1, \cdots, w$, so aus, daß die k Formen

$$(48) \qquad \psi_1, \cdots, \psi_v \text{ und } \phi_1, \cdots, \phi_w$$

linear unabhängig sind.

Man hat nun die Überlegung der Einleitung, die zu (2) führte, in etwas verallgemeinerter Form zu wiederholen. Dem Potenzprodukt $J_0^0 J_0'^0 = 1$ sei t_1 zugeordnet. Die Determinante des Systems der k Formen (48) sei Δ, die Unterdeterminanten der Elemente der ersten Spalte von Δ seien $\Gamma_1, \cdots, \Gamma_v, B_1, \cdots, B_w$. Dann ist

$$(49) \qquad \Delta = \Gamma_1\psi_1 + \cdots + \Gamma_v\psi_v + B_1\phi_1 + \cdots + B_w\phi_w.$$

Aus Satz 7 hatte sich ergeben, daß die ganzen rationalen Koeffizienten des Polynoms $\sigma_{ab}(x)$ für $a = 0, \cdots, n+k^2-1$ und $b = 1, \cdots, k$ die Größenordnung $(n!)^{3+\varepsilon}$ haben. Das Polynom σ_{ab} ist vom Grade $2n+a-1 \leq 3n+k^2-2$. Die Determinante Δ ist ein Polynom in ξ vom Grade $w(3n+k^2-2)$, dessen Koeffizienten ganz rational sind und in n und G die Größenordnung $(n!)^{3w+\varepsilon}G^v$ besitzen. Bei festem ξ sind ferner die Unterdeterminanten $\Gamma_1, \cdots, \Gamma_v$ von der Größenordnung $(n!)^{3w+\varepsilon}G^{v-1}$ und die Unterdeterminanten $B_1, \cdots, B_w$ von der Größenordnung $(n!)^{3(w-1)+\varepsilon}G^v$. Da $\phi_1, \cdots, \phi_w$ die mit $n!$ multiplizierte Majorante (41) haben, so sind für festes ξ die Zahlen $\phi_1, \cdots, \phi_w$ von der Größenordnung $(n!)^{3+\varepsilon-(2k-1)}$.

Die rechte Seite von (49) ist daher

$$(50) \qquad O\left((n!)^{3w+\varepsilon}G^v\left\{\frac{|g(J_0(\xi), J_0'(\xi))|}{G} + (n!)^{1-2k}\right\}\right).$$

Nun sei speziell ξ eine algebraische Zahl vom Grade m. Man wähle eine natürliche Zahl c derart, daß $c\xi$ ganz ist. Dann ist $c^{w(3n+k^2-2)}\Delta$ eine ganze Zahl aus dem

Körper von ξ; da sie von 0 verschieden ist, so ist ihre Norm absolut genommen ≥ 1. Mit Rücksicht auf die Abschätzung der Koeffizienten von Δ folgt daher aus (49) und (50)

$$(51) \qquad 1 < K G^{vm} (n!)^{3wm+\varepsilon} \left\{ \frac{|g(J_0(\xi), J_0'(\xi))|}{G} + (n!)^{1-2k} \right\},$$

wo K nicht von n abhängt und ≥ 1 ist. Man wähle jetzt $\varepsilon = 1$ und $r = 4pm$, dann ist wegen $l = p + r$

$$2k - 2 = (p + r + 1)(p + r + 2) - 2 > p^2(4m + 1)^2 > 8p^2 m(2m + 1),$$
$$v = \tfrac{1}{2}(r + 1)(r + 2) < 2p^2(2m + 1)^2,$$
$$w = \tfrac{1}{2}(p + r + 1)(p + r + 2) - \tfrac{1}{2}(r + 1)(r + 2) = \tfrac{1}{2}p(p + 2r + 3) \leq 2p^2(2m + 1).$$

Man wähle ferner für n die *kleinste* natürliche Zahl, die (42) und der Bedingung

$$n! > 2KG^{2m+1}$$

genügt. Dann ist

$$(n!)^{2k-1} > (n!)^{1+8p^2 m(2m+1)} > (n!)^{1+6p^2 m(2m+1)}\, 2KG^{2p^2 m(2m+1)^2} \geq 2KG^{vm}(n!)^{3wm+\varepsilon},$$

also nach (51)

$$|g(J_0(\xi), J_0'(\xi))| > G(n!)^{1-2k}.$$

Beachtet man, daß $n!$ die Größenordnung $G^{2m+1}\log G$ besitzt und

$$(2k - 1)(2m + 1) \leq 3m\{(4pm + p + 1)(4pm + p + 2) - 1\} \leq 123\, p^2 m^3$$

ist, so folgt der

Hauptsatz:

Es sei ξ eine von 0 verschiedene algebraische Zahl m-ten Grades. Es sei $g(y, z)$ ein Polynom von der Dimension p in y und z, dessen Koeffizienten ganz rational, nicht sämtlich 0 und absolut $\leq G$ sind. Dann gilt für eine gewisse nur von ξ und p abhängige positive Zahl c die Ungleichung

$$(52) \qquad |g(J_0(\xi), J_0'(\xi))| > cG^{-123p^2 m^3}.$$

Insbesondere besteht also zwischen $J_0(\xi)$ und $J_0'(\xi)$ keine algebraische Gleichung mit rationalen Koeffizienten, und, spezieller, die Zahl $J_0(\xi)$ ist transzendent.

Durch die im Hauptsatz gewählte Formulierung hat die negative Aussage, daß die Zahlen $J_0(\xi)$ und $J_0'(\xi)$ für algebraisches $\xi \neq 0$ algebraisch unabhängig sind, eine positive Wendung bekommen; es wird nämlich eine positive Schranke für den Abstand des Wertes eines beliebigen mit rationalen Koeffizienten aus $J_0'(\xi)$ und $J_0(\xi)$ gebildeten Polynoms von 0 angegeben. Vermöge dieser Abschätzung läßt sich mit den transzendenten Zahlen $J_0(\xi)$, $J_0'(\xi)$ in derselben Weise wirklich *rechnen* wie mit den algebraischen Zahlen; denn man kann ja *entscheiden*, wie ein gegebener algebraischer Ausdruck in J_0, J_0' mit algebraischen Koeffizienten zu einer gegebenen rationalen Zahl gelegen ist.

Für die Konstante c in (52) ließe sich leicht ein expliziter Ausdruck als Funktion von ξ und p angeben, und der Exponent $123\, p^2 m^3$ läßt sich auch noch verkleinern, indem man die Abschätzungen schärfer ausführt. Im Falle $p = 1$ läßt sich der »genaue« Exponent ermitteln. Der so entstehende arithmetische Satz hat dann wieder ein algebraisches Analogon, worüber noch kurz berichtet werden möge.

Es gibt drei Polynome n-ten Grades, etwa $f(x)$, $g(x)$, $h(x)$, so daß die Potenzreihe für

$$f(x)J_0(x) + g(x)J_0'(x) + h(x) = R(x)$$

mit der Potenz x^{3n+2} oder einer noch höheren Potenz beginnt und f, g, h nicht identisch gleich 0 sind. Dann ist

$$(f' - g)J_0 + \left(g' + f - \frac{g}{x}\right)J_0' + h' = R'$$

$$\left(f'' - f - 2g' + \frac{g}{x}\right)J_0 + \left(2f' - \frac{f}{x} - g + g'' - \frac{2g'}{x} + \frac{2g}{x^2}\right)J_0' + h'' = R'',$$

wofür kurz

$$f_1 J_0 + g_1 J_0' + h_1 = R'$$
$$f_2 J_0 + g_2 J_0' + h_2 = R''$$

geschrieben werden möge. Da J_0 gerade, J_0' ungerade ist, so können die Polynome f, h entweder beide gerade oder beide ungerade gewählt werden, in letzterem Falle ist g gerade. In beiden Fällen ist fg durch x teilbar. Daher sind die Ausdrücke

$$fg_1 - gf_1, \quad x(f_2 g - g_2 f), \quad x^2(f_1 g_2 - g_1 f_2)$$

Polynome. Die Determinante

$$(53) \qquad \begin{vmatrix} f & g & h \\ f_1 & g_1 & h_1 \\ f_2 & g_2 & h_2 \end{vmatrix} = (f_1 g_2 - g_1 f_2) R + (f_2 g - g_2 f) R' + (fg_1 - gf_1) R''$$

ist also einerseits eine durch x^{3n} teilbare ganze Funktion, andererseits eine rationale Funktion vom Grade $3n$. Sie hat daher den Wert γx^{3n}, und die Konstante γ ist nach Satz 4 von 0 verschieden. Aus (53) folgt, daß R wirklich mit der $(3n+2)$-ten Potenz von x beginnt und nicht mit einer höheren. Demnach sind die Polynome f, g, h bis auf einen gemeinsamen konstanten Faktor eindeutig bestimmt. Es ist ferner

$$J_0 : J_0' : 1 = (gh_1 - hg_1 + Rg_1 - R'g) : (hf_1 - fh_1 - Rf_1 + R'f) : (fg_1 - gf_1),$$

und dies liefert eine Approximation von J_0 und J_0' durch rationale Funktionen mit demselben Nenner. Vielleicht dürften die hier auftauchenden verallgemeinerten Kettenbruchentwicklungen auch bei andern linearen Differentialgleichungen von Bedeutung sein.

Der Satz, daß der Ausdruck $fJ_0 + gJ_0' + h$ zwar durch x^{3n+2}, aber nicht x^{3n+3} teilbar sein kann, wenn f, g, h Polynome n-ten Grades bedeuten, hat folgendes arithmetische Analogon:

Es sei r eine von 0 verschiedene rationale Zahl. Es seien a, b, c drei ganze rationale Zahlen, deren absolute Beträge das positive Maximum M haben. Dann ist

$$(54) \qquad |aJ_0(r) + bJ_0'(r) + c| > c_1 M^{-2-\varepsilon},$$

wo ε eine beliebige positive Zahl bedeutet und $c_1 > 0$ nur von r und ε abhängt.

Hier ist der Exponent $-2-\varepsilon$ bis auf das beliebig kleine ε der günstigste; denn für irgendwelche reellen Zahlen ρ, ς, τ hat stets die Ungleichung $|a\rho + b\varsigma + c\tau| \leq (|\rho| + |\varsigma| + |\tau|) M^{-2}$ unendlich viele Lösungen in ganzen rationalen Zahlen a, b, c.

Der Beweis von (54) ergibt sich auf genau demselben Wege, der zu (52) geführt hat, wenn man die Abschätzungen etwas verfeinert. Ebenso folgt allgemeiner:

Es sei $r \neq 0$ rational. Man bilde ein Polynom P in $J_0(r)$, $J_0'(r)$ von der Dimension p. Seine $\frac{1}{2}(p+1)(p+2) = q$ Koeffizienten seien ganz rational und absolut $\leq M$. Dann ist

$$(55) \qquad |P(J_0(r), J_0'(r))| > c_2 M^{1-q-\varepsilon}$$

für beliebiges $\varepsilon > 0$ und nur von r, p, ε abhängiges $c_2 > 0$.

Und andererseits kann man die Koeffizienten von P auf unendlich viele Weisen so bestimmen, daß

$$|P(J_0(r), J_0'(r))| < c_3 M^{1-q}$$

ist, für ein gewisses nur von r und p abhängiges c_3. Die Ungleichung (55) bringt also zum Ausdruck, wie sehr sich die Werte $J_0(r)$ und $J_0'(r)$ dagegen sträuben, durch eine algebraische Gleichung verbunden zu werden.

Genau wie im früher behandelten Fall $p = 1$ gibt es ein algebraisches Analogon zu (55), welches zeigt, wie sehr sich die Funktionen $J_0(x)$ und $J_0'(x)$ dagegen sträuben, durch eine algebraische Gleichung, deren Koeffizienten Polynome in x sind, miteinander verknüpft zu werden; dies bildet dann eine Verfeinerung des Satzes 2, daß $J_0(x)$ keiner algebraischen Differentialgleichung erster Ordnung genügt.

§ 4. Weitere Anwendungen der Methode.

I.

Die E-Funktion

$$K_\lambda(x) = \Gamma(\lambda+1)\left(\frac{x}{2}\right)^{-\lambda} J_\lambda(x) = \sum_{n=0}^{\infty} \frac{(-1)^n}{n!\,(\lambda+1)\cdots(\lambda+n)}\left(\frac{x}{2}\right)^{2n}$$

mit rationalem $\lambda \neq -1, -2, \cdots$ läßt sich genau so untersuchen, wie es mit $K_0 = J_0$ geschehen ist. Man erhält auch hier das Resultat, daß für kein algebraisches $\xi \neq 0$ zwischen den Zahlen $K_\lambda(\xi)$ und $K'_\lambda(\xi)$ eine algebraische Gleichung mit rationalen Koeffizienten besteht; auszunehmen ist nur der Fall, daß λ die Hälfte einer ungeraden Zahl ist. Insbesondere sind also die von 0 verschiedenen Nullstellen der Besselschen Funktionen $J_\lambda(x)$ für rationales λ stets transzendent, und dies gilt auf Grund des Lindemannschen Satzes auch noch für ungerades 2λ.

Aus den bekannten Relationen

(56) $$J'_\lambda = \frac{\lambda}{x} J_\lambda - J_{\lambda+1},$$

$$\frac{J_{\lambda-1}}{J_\lambda} = \frac{2\lambda}{x} - \cfrac{1}{\cfrac{2\lambda+2}{x} - \cdots}$$

folgt, daß der Kettenbruch

$$i\,\frac{J_{\lambda-1}(2ix)}{J_\lambda(2ix)} = \frac{\lambda}{x} + \cfrac{1}{\cfrac{\lambda+1}{x} + \cfrac{1}{\cfrac{\lambda+2}{x} + \cdots}}$$

für rationales λ und algebraisches $x \neq 0$ eine transzendente Zahl darstellt. Für ungerades 2λ ist dies im Satz von Lindemann enthalten.

Daraus ergibt sich die Transzendenz des Kettenbruchs

$$r_1 + \cfrac{1}{r_2 + \cfrac{1}{r_3 + \cdots}},$$

falls $r_1, r_2, r_3, \cdots$ rationale Zahlen sind, die eine arithmetische Reihe erster Ordnung bilden.

Speziell ist der Kettenbruch

$$1 + \cfrac{1}{2 + \cfrac{1}{3 + \cdots}}$$

transzendent.

II.

Bekanntlich haben $J_\lambda(x)$ und $J_{\lambda+1}(x)$ keine gemeinsame von 0 verschiedene Nullstelle, wie sich aus (56) ergibt. Es soll nun gezeigt werden, daß auch J_λ und $J_{\lambda+n}$ für $n = 2, 3, \cdots$ keine gemeinsame Nullstelle $\neq 0$ haben, falls λ rational ist und für negatives ganzes rationales λ der Wert $n = -2\lambda$ ausgelassen wird. Nach (56) ist nämlich

$$J_{\lambda+n} = P J_\lambda + Q J'_\lambda,$$

wo P und Q rationale Funktionen von x mit rationalen Koeffizienten bedeuten. Wäre Q identisch gleich 0, so folgte mit $\lambda + n = \mu$

$$\frac{d^2(PJ_\lambda)}{dx^2} + \frac{1}{x}\frac{d(PJ_\lambda)}{dx} + \left(1 - \frac{\mu^2}{x^2}\right) P J_\lambda = 0,$$

also

(57) $$\left(P'' + \frac{P'}{x} + \frac{\lambda^2-\mu^2}{x^2} P\right) J_\lambda + 2P' J'_\lambda = 0.$$

Ist nun zunächst 2λ keine ungerade Zahl, so folgte nach Satz 2

$$P' = 0, \quad \lambda^2 = \mu^2, \quad n = -2\lambda, \quad \lambda \text{ ganz},$$

und dies ist der triviale Ausnahmefall $(-1)^\lambda J_\lambda = J_{-\lambda}$. Ist aber 2λ ungerade, so gilt bekanntlich

$$J_\lambda = a e^{ix} + b e^{-ix} \tag{58}$$

mit rationalen Funktionen a und b, von denen keine identisch 0 ist, also

$$J'_\lambda = (a' + ia) e^{ix} + (b' - ib) e^{-ix}, \tag{59}$$

und es ist die Funktion $(a' + ia)b - a(b' - ib) = 2iab + ba' - ab'$ nicht identisch 0, weil sie denselben Grad hat wie ab; aus (57), (58) und (59) folgt wieder $P' = 0$, $\lambda^2 = \mu^2$; es wären also $J_{-\lambda}$ und J_λ proportional, was ein Widerspruch ist.

Daher ist Q nicht identisch gleich 0. Ist nun $\alpha \neq 0$ eine Nullstelle von J_λ, so ist dies eine einfache Nullstelle, also $J'_\lambda(\alpha) \neq 0$. Da α transzendent ist, so ist $P(\alpha)$ endlich und $Q(\alpha) \neq 0$, also

$$J_{\lambda+n}(\alpha) = Q(\alpha) J'_\lambda(\alpha) \neq 0.$$

III.

Einfacher als $J_0(x)$ läßt sich die Lösung von

$$y' + \left(\frac{\lambda}{x} - 1\right) y = \frac{\lambda}{x} \qquad (\lambda \neq -1, -2, \cdots) \tag{60}$$

behandeln, nämlich

$$y = x^{-\lambda} e^x \left(\int_0^x \lambda t^{\lambda-1} e^{-t} dt + c \right). \tag{61}$$

Für den Wert 0 der Integrationskonstanten c hat man

$$y_0 = 1 + \frac{x}{\lambda+1} + \frac{x^2}{(\lambda+1)(\lambda+2)} + \cdots.$$

Es ergibt sich die Transzendenz von y_0 für jedes rationale λ und jedes algebraische $x \neq 0$. Folglich sind die Nullstellen der Funktion

$$\int_0^1 t^{\lambda-1} e^{-tx} dt \tag{62}$$

transzendent. Dies enthält für $\lambda = 1$ die Transzendenz von π, für $x = 1$ die Irrationalität der Nullstellen der »unvollständigen« Gammafunktion

$$\int_0^1 t^{x-1} e^{-t} dt.$$

Für algebraisches c und rationales λ ist die rechte Seite von (61) ebenfalls eine E-Funktion. Also ist der Ausdruck (62) für rationales λ und algebraisches $x \neq 0$ eine transzendente Zahl, also auch

$$\int_0^x e^{-t^\lambda} dt.$$

IV.

Die Lösung von (60) ist enthalten in der Lösung von

$$xy'' + (\lambda - x) y' - \varkappa y = 0;$$

hiervon ist ein Integral

$$y = 1 + \frac{\varkappa}{\lambda} \frac{x}{1!} + \frac{\varkappa(\varkappa+1)}{\lambda(\lambda+1)} \frac{x^2}{2!} + \cdots = \int_0^1 t^{\varkappa-1}(1-t)^{\lambda-\varkappa-1} e^{tx} dt : \int_0^1 t^{\varkappa-1}(1-t)^{\lambda-\varkappa-1} dt.$$

Wenn $\varkappa$ und λ rational sind, läßt sich diese Funktion in derselben Weise wie $J_0(x)$ untersuchen.

V.

Durch leichte Verallgemeinerung der früheren Überlegungen ist es möglich, den folgenden Satz zu beweisen, welcher sowohl den Hauptsatz über $J_0(x)$ als auch den allgemeinen LINDEMANNschen Satz enthält:

Es seien $\xi, \alpha_1, \cdots, \alpha_n$ algebraische Zahlen, und zwar seien $\alpha_1, \cdots, \alpha_n$ voneinander und ξ von 0 verschieden. Es seien $P_1(x, y), \cdots, P_n(x, y)$ Polynome mit algebraischen Koeffizienten, die nicht sämtlich 0 sind. Dann ist die Zahl

$$(63) \qquad P_1(J_0(\xi),\ J_0'(\xi))e^{\alpha_1} + \cdots + P_n(J_0(\xi),\ J_0'(\xi))e^{\alpha_n} \neq 0.$$

Anders ausgedrückt: Zwischen $J_0(\xi)$, $J_0'(\xi)$ und Werten der Exponentialfunktion für algebraisches Argument besteht keine nichttriviale algebraische Relation. Speziell ist also $J_0(\xi)$ transzendent in bezug auf den Körper von e.

Zum Beweise bedarf man einer Verallgemeinerung des Satzes 2; es ist zu zeigen, daß die Funktion

$$P_1(J_0(\xi x),\ J_0'(\xi x))e^{\alpha_1 x} + \cdots + P_n(J_0(\xi x),\ J_0'(\xi x))e^{\alpha_n x}$$

nicht identisch in x verschwindet. Dies geschieht nach der Methode des Beweises von Satz 2; vgl. VI.

Es ergibt sich auch eine explizite positive untere Schranke für den absoluten Betrag der linken Seite von (63). Insbesondere gilt für den Spezialfall des HERMITEschen Satzes:

Es seien $m_0, \cdots, m_n$ ganze rationale Zahlen, deren absolute Beträge das Maximum $M > 0$ haben. Dann ist

$$|m_0 e^n + m_1 e^{n-1} + \cdots + m_{n-1} e + m_n| > c_4 M^{-n-\varepsilon},$$

wo $c_4 > 0$ ist und nur von ε und n abhängt. Andererseits ist wieder

$$|m_0 e^n + \cdots + m_n| < c_5 M^{-n}$$

für ein geeignetes c_5 in unendlich vielen ganzen rationalen Zahlen $m_0, \cdots, m_n$ lösbar.

Analog gilt der algebraische Satz:

Sind $f_0(x), \cdots, f_n(x)$ Polynome h-ten Grades, die nicht sämtlich identisch gleich 0 sind, so verschwindet die Funktion

$$f_0(x)e^{nx} + f_1(x)e^{(n-1)x} + \cdots + f_{n-1}(x)e^x + f_n(x)$$

bei $x = 0$ höchstens von der Ordnung $(n+1)(h+1) - 1$, und diese Ordnung wird, was ja trivial ist, durch geeignete Polynome $f_0, \cdots, f_n$ auch wirklich erreicht.

Für $n = 1$ folgen beide Sätze aus dem Kettenbruch für e^x. Die Sätze bringen zum Ausdruck, wie sehr sich die Zahl e und die Funktion e^x dagegen sträuben, algebraisch zu sein.

Für π gewinnt man die Ungleichung

$$|m_0 \pi^n + m_1 \pi^{n-1} + \cdots + m_n| > c_6 M^{-M^\varepsilon}$$

für beliebig kleines positives ε und nur von ε und n abhängiges positives c_6.

VI.

Die allgemeinere Frage nach der algebraischen Unabhängigkeit der Werte von $J_0(\xi)$ für verschiedene algebraische Werte von $\xi^2 \neq 0$ läßt sich ebenfalls beantworten, so daß man über $J_0(x)$ ebensogut informiert ist wie auf Grund des LINDEMANNschen Satzes über $\exp(x)$. Es läßt sich nämlich zeigen:

Es seien $\xi_1^2, \cdots, \xi_n^2$ voneinander und von 0 verschiedene algebraische Zahlen. Dann sind die $2n$ Zahlen $J_0(\xi_1), J_0'(\xi_1), \cdots, J_0(\xi_n); J_0'(\xi_n)$ voneinander algebraisch unabhängig im Körper der rationalen Zahlen.

Allgemeiner gilt dieser Satz auch für K_λ an Stelle von J_0, wenn λ eine rationale Zahl bedeutet, die nicht die Hälfte einer ungeraden Zahl ist. Und auch das läßt sich weiter verallgemeinern, indem man verschiedene Werte von λ simultan betrachtet. Man erhält so folgende Aussage:

Es seien $\lambda_1, \cdots, \lambda_m$ rationale Zahlen; keine der Zahlen $2\lambda_1, \cdots, 2\lambda_m$ sei ungerade, keine der Summen $\lambda_k + \lambda_l$ und der Differenzen $\lambda_k - \lambda_l$ ($k = 1, \cdots, m$; $l = 1, \cdots, m$; $k \neq l$) sei ganz. Es seien $\xi_1^2, \cdots, \xi_n^2$ voneinander und von 0 verschiedene algebraische Zahlen. Dann besteht zwischen den $2mn$ Größen $K_\lambda(\xi), K'_\lambda(\xi)$ ($\lambda = \lambda_1, \lambda_2, \cdots, \lambda_m$; $\xi = \xi_1, \xi_2, \cdots, \xi_n$) keine algebraische Gleichung mit rationalen Koeffizienten.

Der Beweis verläuft entsprechend wie in dem ausführlich behandelten Fall $m = 1$, $\lambda_1 = 0$, $n = 1$. Es genügt, den algebraischen Teil des Beweises zu behandeln, der zur Verallgemeinerung des Satzes 3 führt, nämlich zu

Satz 9:

Es seien $\lambda_1, \cdots, \lambda_m$ irgendwelche Zahlen, von denen keine die Hälfte einer ungeraden Zahl ist und kein Paar eine ganze rationale Summe oder Differenz besitzt. Ferner seien die Zahlen $\xi_1^2, \cdots, \xi_n^2$ voneinander und von 0 verschieden. Es seien J_λ und Y_λ voneinander linear unabhängige Lösungen der Bessel*schen Differentialgleichung. Dann sind die $3mn$ Funktionen $J_\lambda(\xi x)$, $J'_\lambda(\xi x)$, $Y_\lambda(\xi x)$ ($\lambda = \lambda_1, \cdots, \lambda_m$; $\xi = \xi_1, \cdots, \xi_n$) algebraisch unabhängig im Körper der rationalen Funktionen von x.*

Beweis: Die Besselschen Funktionen mit dem Argument ξx genügen der Differentialgleichung

$$(64) \qquad y'' + \frac{1}{x}y' + \left(\xi^2 - \frac{\lambda^2}{x^2}\right)y = 0.$$

Es seien u und v zwei linear unabhängige Lösungen. Die zu den mn Paaren ξ, λ gehörigen Lösungen seien in irgendeiner Reihenfolge mit $u_1, v_1; u_2, v_2; \cdots$ bezeichnet. Man hat zu beweisen, daß zwischen den $3mn+1$ Funktionen $x, u_1, u'_1, v_1, u_2, u'_2, v_2, \cdots$ keine algebraische Gleichung mit konstanten Koeffizienten besteht. Für $m = 1$, $n = 1$ ist dies durch Satz 3 erledigt. Es sei $mn > 1$ und die Anzahl r der in der Gleichung wirklich vorkommenden Tripel u, u', v möglichst klein; ferner sei bei diesem r die Anzahl s der wirklich auftretenden Funktionen v möglichst klein. Man hat also etwa eine Gleichung zwischen $u_1, u'_1, u_2, u'_2, \cdots, u_r, u'_r$ und $v_1, \cdots, v_s$ mit $s \leq r$, deren Koeffizienten Polynome in x sind; und in jeder solchen Gleichung muß eine der Funktionen u_r, u'_r, v_r und im Falle $s > 0$ die Funktion v_s wirklich auftreten.

Zunächst wird gezeigt, daß s den Wert 0 hat. Man hat nur die Überlegungen vom Beweise des Satzes 3 zu wiederholen; an die Stelle der Gleichung (18) tritt dann eine Gleichung zwischen $x, u_2, u'_2, \cdots, u_r, u'_r$ und $v_2, \cdots, v_s$ und $\lambda u_1 + \mu v_1$, $\lambda u'_1 + \mu v'_1$, wo λ und μ geeignete Konstanten bedeuten. Ersetzt man dann $\lambda u_1 + \mu v_1$ durch u_1, so kommt man zu einer Gleichung mit demselben r und kleinerem s. Daher ist $s = 0$.

Nun soll nachgewiesen werden, daß auch eine Gleichung existiert, in der u'_1 nicht auftritt. Man schließt wie beim Beweise von Satz 2. Ist $\Phi = 0$ eine u'_1 enthaltende irreduzible algebraische Gleichung zwischen $x, u_1, u'_1, \cdots, u_r, u'_r$, so kann man aus $\Phi = 0$ und $\frac{d\Phi}{dx} = 0$ unter Benutzung von (64) entweder u'_1 eliminieren, oder man erhält analog zu (11) eine Gleichung

$$(65) \qquad Q(\lambda_1 u_1 + \mu_1 v_1, \lambda_1 u'_1 + \mu_1 v'_1, \lambda_2 u_2 + \mu_2 v_2, \cdots) = k(\lambda_1, \mu_1, \lambda_2, \mu_2, \cdots) x^b e^{ax - \frac{c}{x}};$$

hierin bedeutet k ein Polynom mit konstanten Koeffizienten, das in jedem der r Paare $\lambda_a, \mu_a (a = 1, \cdots, r)$ homogen ist, und $Q(u_1, u'_1, \cdots)$ ein Polynom in $x, u_1, u'_1, \cdots$, das in jedem der Paare $u_a, u'_a (a = 1, \cdots, r)$ homogen ist. Man setze noch $\lambda_a : \mu_a = \rho_a$ und unterwerfe diese Parameter der Bedingung $k = 0$, dann ist $\rho_1 = \rho$ eine algebraische Funktion von $\rho_2, \cdots, \rho_r$. Wäre ρ identisch konstant, so würde für diesen Wert von $\rho = \lambda_1 : \mu_1$ die linke Seite von (65) identisch in $\lambda_2, \mu_2, \cdots, \lambda_r, \mu_r$ verschwinden; dann wäre aber der Koeffizient jedes Potenzproduktes von $u_2, u'_2, \cdots, u_r, u'_r$ in $Q(\lambda_1 u_1 + \mu_1 v_1, \lambda_1 u'_1 + \mu_1 v'_1, u_2, u'_2, \cdots, u_r, u'_r)$ gleich 0, und dies lieferte eine Gleichung zwischen $\lambda_1 u_1 + \mu_1 v_1, \lambda_1 u'_1 + \mu v'_1, x$, gegen Satz 3. Also ist ρ nicht konstant bei variabeln $\rho_2, \cdots, \rho_r$.

Die Funktion $\frac{\rho u_1' + v_1'}{\rho u_1 + v_1}$ ist eine algebraische Funktion von x und den $r-1$ Funktionen $\frac{\rho_2 u_2' + v_2'}{\rho_2 u_2 + v_2}, \cdots, \frac{\rho_r u_r' + v_r'}{\rho_r u_r + v_r}$, die mit A bezeichnet werden möge. Man wähle für $\rho_2, \cdots, \rho_r$ vier beliebige Wertsysteme, die durch die oberen Indizes I, $\cdots$, IV bezeichnet werden mögen; die entsprechenden Indizes werden an die zugehörigen Werte von ρ und A angeheftet. Aus der Gleichung

$$\text{(66)} \qquad \frac{\rho u_1' + v_1'}{\rho u_1 + v_1} = A$$

folgt dann die Gleichheit der Doppelverhältnisse

$$\text{(67)} \qquad \frac{\rho^{\mathrm{I}} - \rho^{\mathrm{III}}}{\rho^{\mathrm{I}} - \rho^{\mathrm{IV}}} : \frac{\rho^{\mathrm{II}} - \rho^{\mathrm{III}}}{\rho^{\mathrm{II}} - \rho^{\mathrm{IV}}} = \frac{A^{\mathrm{I}} - A^{\mathrm{III}}}{A^{\mathrm{I}} - A^{\mathrm{IV}}} : \frac{A^{\mathrm{II}} - A^{\mathrm{III}}}{A^{\mathrm{II}} - A^{\mathrm{IV}}},$$

und zwar gilt dies identisch in den $4(r-1)$ Variabeln $\rho_2^{\mathrm{I}}, \cdots, \rho_r^{\mathrm{I}}, \cdots, \rho_2^{\mathrm{IV}}, \cdots, \rho_r^{\mathrm{IV}}$, wenn unter ρ die k zu 0 machende algebraische Funktion von $\rho_2, \cdots, \rho_r$ verstanden wird. Nach (14) ist für ein konstantes $a \neq 0$

$$\text{(68)} \qquad u'v - v'u = \frac{a}{x};$$

setzt man die hieraus für $v_2', \cdots, v_r'$ folgenden Werte in (67) ein, so entsteht eine algebraische Gleichung zwischen x und nur $r-1$ Tripeln $u_2, u_2', v_2, \cdots, u_r, u_r', v_r$, also eine Gleichung, die identisch in $u_2, u_2', v_2, \cdots, u_r, u_r', v_r$ gilt. Nun hängt aber A nur von den Verhältnissen $\frac{\rho_2 u_2' + v_2'}{\rho_2 u_2 + v_2}, \cdots$ ab, also auch nur von den Verhältnissen $u_2' : v_2' : u_2 : v_2, \cdots$; bei willkürlichen $u' : v' : u : v$ bedeutet (68) die Normierung eines Proportionalitätsfaktors von u', v', u, v, und folglich gilt (67) sogar identisch in den $4(r-1)$ Variabeln $u_2, u_2', v_2, v_2', \cdots$. Man kann daher die in ρ_2 lineare Funktion $\frac{\rho_2 u_2' + v_2'}{\rho_2 u_2 + v_2}$ durch

$$\frac{\rho_2 - \rho_2^{\mathrm{III}}}{\rho_2 - \rho_2^{\mathrm{IV}}} : \frac{\rho_2^{\mathrm{II}} - \rho_2^{\mathrm{III}}}{\rho_2^{\mathrm{II}} - \rho_2^{\mathrm{IV}}}$$

ersetzen und analog für die andern Indizes $3, \cdots, r$. Dann werden die Argumente von A^{II} sämtlich gleich 1, die von A^{III} gleich 0, die von A^{IV} gleich ∞, und die Argumente von A^{I} sind die Doppelverhältnisse von $\rho_a^{\mathrm{I}}, \rho_a^{\mathrm{II}}, \rho_a^{\mathrm{III}}, \rho_a^{\mathrm{IV}}$ für $a = 2, \cdots, r$. Nach (67) ist dann das Doppelverhältnis von $\rho^{\mathrm{I}}, \rho^{\mathrm{II}}, \rho^{\mathrm{III}}, \rho^{\mathrm{IV}}$ eine algebraische Funktion dieser $r-1$ Doppelverhältnisse, und andererseits ist ρ^{I} eine algebraische Funktion von $\rho_2^{\mathrm{I}}, \cdots, \rho_r^{\mathrm{I}}$ allein, $\cdots$, ρ^{IV} dieselbe algebraische Funktion von $\rho_2^{\mathrm{IV}}, \cdots, \rho_r^{\mathrm{IV}}$ allein.

Ist nun für eine in einem Intervall differentiierbare und nicht identisch konstante Funktion $f(x)$ das Doppelverhältnis $\Delta = \frac{f(x_1) - f(x_3)}{f(x_1) - f(x_4)} : \frac{f(x_2) - f(x_3)}{f(x_2) - f(x_4)}$ eine Funktion von $\frac{x_1 - x_3}{x_1 - x_4} : \frac{x_2 - x_3}{x_2 - x_4} = D$, so folgt, indem man x_1 und x_2 bei festgehaltenem Wert von D nach x_3 konvergieren läßt, die Relation

$$\Delta : D = \lim \frac{f(x_1) - f(x_3)}{x_1 - x_3} : \frac{f(x_2) - f(x_3)}{x_2 - x_3} = 1;$$

also $\Delta = D$.

In (67) ist also die rechte Seite gleich dem ersten Argument von A^{I}, ferner sind $A^{\mathrm{II}}, A^{\mathrm{III}}, A^{\mathrm{IV}}$ algebraische Funktionen von x. Folglich ist A eine lineare Funktion von $\frac{\rho_2 u_2' + v_2'}{\rho_2 u_2 + v_2}$, deren Koeffizienten algebraische Funktionen von x sind. Nach (66) ist dann ρ eine lineare Funktion von ρ_2. Werden u_1 und v_1 durch geeignete homogene lineare Verbindungen von u_1 und v_1 mit konstanten Koeffizienten ersetzt, so ist $\rho_2 = \rho$. Aus

$$\frac{\rho u_1' + v_1'}{\rho u_1 + v_1} = \frac{\alpha \dfrac{\rho u_2' + v_2'}{\rho u_2 + v_2} + \beta}{\gamma \dfrac{\rho u_2' + v_2'}{\rho u_2 + v_2} + \delta}$$

mit algebraischen α, β, γ, δ folgt nun aber, wenn nach ρ differentiiert und $\rho = \infty$ gesetzt wird,

$$\frac{u_1' v_1 - v_1' u_1}{u_1^2} = (\alpha \delta - \beta \gamma) \frac{u_2' v_2 - v_2' u_2}{(\gamma u_2 + \delta u_2)^2},$$

also nach (68)

$$u_r = p u_2 + q u_2'$$

mit algebraischen Funktionen p und q von x allein. Diese Gleichung ist von u_1' frei.

Es gibt also auch eine Gleichung zwischen x, u_2, u_2', $\cdots$, u_r, u_r' und u_1. Sie heiße $\psi = 0$ und sei irreduzibel. Aus $\dfrac{d\psi}{dx} = 0$ ergibt sich dann u_1' als rationale Funktion von x, u_2, u_2', $\cdots$, u_r, u_r' und u_1, und durch Differentiation u_1'' als ebensolche Funktion. Trägt man u_1 an Stelle von y in die linke Seite von (64) ein, so muß der entstehende Ausdruck verschwinden; dies ist ebenfalls eine Gleichung $\chi = 0$ zwischen x, u_2, u_2', $\cdots$, u_r, u_r' und u_1. Die Elimination von u_1 aus $\chi = 0$ und $\psi = 0$ ist unmöglich. Folglich ist die Gleichung $\chi = 0$ erfüllt, falls u_2, $\cdots$, u_r durch beliebige Lösungen der für diese Funktionen geltenden Differentialgleichungen ersetzt werden und u_1 auf Grund von $\psi = 0$ bestimmt wird. Schreibt man also $\lambda_2 u_2 + \mu_2 v_2$, $\lambda_2 u_2' + \mu_2 v_2'$, $\lambda_3 u_3 + \mu_3 v_3$, $\lambda_3 u_3' + \mu_3 v_3'$, $\cdots$ an Stelle von u_2, u_2', u_3, u_3', $\cdots$, so genügt $\lambda_1 u_1 + \mu_1 v_1$ der für u_1 geltenden Gleichung $\psi = 0$; dabei sind λ_1 und μ_1 Funktionen von λ_2, μ_2, λ_3, μ_3, $\cdots$. Es sei

$$u_1 = A(u_2, u_2', \cdots),$$

wo A eine algebraische Funktion seiner Argumente bedeutet; dann ist

(69) $$\lambda_1 u_1 + \mu_1 v_1 = A(\lambda_2 u_2 + \mu_2 v_2, \lambda_2 u_2' + \mu_2 v_2', \cdots).$$

Man ersetze speziell μ_2 durch 1 und wähle für λ_2 drei beliebige Werte λ_2^{I}, λ_2^{II}, λ_2^{III}; dadurch gehe A in A^{I}, A^{II}, A^{III} und λ_1, μ_1 in λ_1^{I}, μ_1^{I}, $\cdots$ über. Aus den drei so entstehenden Gleichungen (69) eliminiere man u_1, v_1; dies liefert

(70) $$\rho_1 A^{\mathrm{I}} + \rho_2 A^{\mathrm{II}} + \rho_3 A^{\mathrm{III}} = 0,$$

wo ρ_1, ρ_2, ρ_3 Funktionen von λ_2^{I}, λ_2^{II}, λ_2^{III} sind, aber nicht von x abhängen. Eliminiert man v_2', v_3', $\cdots$ mit Hilfe von (68), so geht (70) in eine Gleichung zwischen u_2, u_2', v_2, $\cdots$. u_r, u_r', v_r, x über, und diese Gleichung muß dann wieder identisch in den $3(r-1) + 1$ genannten Variabeln bestehen. Die beiden ersten Argumente von A lauten, wenn der untere Index 2 der Kürze halber fortgelassen wird, $\lambda u + v = \xi$ und $\dfrac{u'}{u}(\lambda u + v) - \dfrac{a}{xu} = \eta$. Man entwickle $A(\xi, \eta)$ nach Potenzen von $\xi - v = \lambda u$

(71) $$A(\xi, \eta) = A(v, \eta) + A_v(v, \eta)\lambda u + \tfrac{1}{2}A_{vv}(v, \eta)(\lambda u)^2 + \cdots.$$

Ferner entwickle man $A\left(v, -\dfrac{a}{xu}\right)$ nach Potenzen von u; dies sei

(72) $$A\left(v, -\frac{a}{xu}\right) = c_0 u^{r_0} + c_1 u^{r_1} + \cdots, \qquad (r_0 < r_1 < \cdots)$$

und c_0, c_1, $\cdots$ bedeuten Funktionen von v, von denen keine identisch 0 ist. Da (70) identisch in u, u', v, $\cdots$ gilt, so kann speziell $u' = 0$ gesetzt werden; dadurch geht η in $-\dfrac{a}{xu}$ über. Man trage die aus (71) und (72) folgende Entwicklung

$$A\left(\lambda u+v,-\frac{a}{xu}\right)=(c_0u^{r_0}+c_1u^{r_1}+\cdots)+\left(\frac{\partial c_0}{\partial v}u^{r_0}+\frac{\partial c_1}{\partial v}u^{r_1}+\cdots\right)\lambda u+\cdots$$

in (70) ein und lasse u gegen o konvergieren. Es wird

$$c_0(\rho_1+\rho_2+\rho_3)=0\,,$$

also $\rho_1+\rho_2+\rho_3=0$ und $c_1(\rho_1+\rho_2+\rho_3)=0,\cdots$. Bedeutet $\frac{\partial c_k}{\partial v}$ die erste nicht identisch verschwindende Funktion der Reihe $\frac{\partial c_0}{\partial v}, \frac{\partial c_1}{\partial v},\cdots$, so ist ferner

$$\frac{\partial c_k}{\partial v}(\rho_1\lambda^{\mathrm{I}}+\rho_2\lambda^{\mathrm{II}}+\rho_3\lambda^{\mathrm{III}})=0\,,$$

also $\rho_1\lambda^{\mathrm{I}}+\rho_2\lambda^{\mathrm{II}}+\rho_3\lambda^{\mathrm{III}}=0$ und $\frac{\partial c_{k+1}}{\partial v}(\rho_1\lambda^{\mathrm{I}}+\rho_2\lambda^{\mathrm{II}}+\rho_3\lambda^{\mathrm{III}})=0,\cdots$. Bedeutete $\frac{\partial^2 c_l}{\partial v^2}$ die erste nicht identisch verschwindende Funktion der Reihe $\frac{\partial^2 c_k}{\partial v^2}, \frac{\partial^2 c_{k+1}}{\partial v^2},\cdots$, so folgte ebenso $\rho_1(\lambda^{\mathrm{I}})^2+\rho_2(\lambda^{\mathrm{II}})^2+\rho_3(\lambda^{\mathrm{III}})^2=0$, was ein Widerspruch ist, wenn $\lambda^{\mathrm{I}}, \lambda^{\mathrm{II}}, \lambda^{\mathrm{III}}$ voneinander verschieden gewählt werden. Daher sind $c_0, c_1, c_2, \cdots$ in v konstant oder linear. Mit Rücksicht auf (72) folgt

$$A(\xi,\eta)=B(\eta)+\xi C(\eta)\,,$$

wo B und C algebraische Funktionen von η bedeuten. Nun sei wieder u' beliebig, und man entwickle $B(\eta)$ und $C(\eta)$ nach Potenzen von $\frac{u'}{u}(\lambda u+v)=\eta+\frac{a}{xu}$. Dadurch geht (70) über in

$$\rho_1\Big\{B\left(-\frac{a}{xu}\right)+B'\left(-\frac{a}{xu}\right)\frac{u'}{u}(\lambda^{\mathrm{I}}u+v)+\tfrac{1}{2}B''\left(-\frac{a}{xu}\right)\left(\frac{u'}{u}\right)^2(\lambda^{\mathrm{I}}u+v)^2+\cdots$$
$$+(\lambda^{\mathrm{I}}u+v)\left(C\left(-\frac{a}{xu}\right)+C'\left(-\frac{a}{xu}\right)\frac{u'}{u}(\lambda^{\mathrm{I}}u+v)+\cdots\right)\Big\}+\rho_2\{\cdots\}+\rho_3\{\cdots\}=0\,.$$

Dies gilt identisch in u', also ist der Koeffizient von $u':u$

$$\rho_1\left\{B'\left(-\frac{a}{xu}\right)(\lambda^{\mathrm{I}}u+v)+C'\left(-\frac{a}{xu}\right)(\lambda^{\mathrm{I}}u+v)^2\right\}+\rho_2\{\cdots\}+\rho_3\{\cdots\}=0\,,$$

und zwar identisch in v; folglich $C'\left(-\frac{a}{xu}\right)=0$, $C(\eta)$ konstant. Ferner ist auch der Koeffizient von $(u':u)^2$

$$\rho_1B''\left(-\frac{a}{xu}\right)(\lambda^{\mathrm{I}}u+v)^2+\rho_2B''\left(-\frac{a}{xu}\right)(\lambda^{\mathrm{II}}u+v)^2+\rho_3B''\left(-\frac{a}{xu}\right)(\lambda^{\mathrm{III}}u+v)^2=0$$

identisch in v; also $B''\left(-\frac{a}{xu}\right)=0$, $B(\eta)$ linear.

Es hat sich ergeben, daß A eine lineare Funktion seiner beiden ersten Argumente ist; nach (69) sind dann auch λ_1 und μ_1 lineare Funktionen von λ_2 und μ_2. Setzt man $\lambda_2=0$, $\mu_2=0$, so ist (69) eine Gleichung, die das Variabelnpaar u_2, u_2' nicht mehr enthält, es müssen dann also λ_1 und μ_1 verschwinden, und es darf ferner $A(u_2, u_2', \cdots)$ kein von u_2, u_2' freies Glied enthalten. Also ist A homogen linear in u_2, u_2', und λ_1, μ_1 sind homogen linear in λ_2, μ_2. Enthielte A nun noch ein weiteres Variabelnpaar u_3, u_3', so wären λ_1, μ_1 auch homogen linear in λ_3, μ_3, etwa $\lambda_1=a\lambda_3+b\mu_3$, $\mu_1=c\lambda_3+d\mu_3$. Da $ad-bc$ eine homogene quadratische Funktion von λ_2, μ_2 ist, so könnte man für $\lambda_2, \mu_2, \lambda_3, \mu_3$ Werte finden, die λ_1 und μ_1 zu o machen, aber selber nicht sämtlich o sind. Dann wäre aber (69) eine Gleichung, die das Variabelnpaar u_1, u_1' nicht mehr enthält.

Man ist somit auf eine Gleichung

$$u_1=\alpha u_2+\beta u_2'$$

geführt worden, wo α und β algebraische Funktionen von x sind. Trägt man diesen Wert von u_1 in die Differentialgleichung ein und benutzt die Differentialgleichung für u_2 zur Elimination von u_2'' und u_2''', so erhält man

$$Ru_2' + Su_2 = 0$$

mit

$$R = \beta'' - \frac{1}{x}\beta' + \left(\xi_1^2 - \xi_2^2 + \frac{\lambda_2^2 - \lambda_1^2 + 1}{x^2}\right)\beta + 2\alpha'$$

$$S = \alpha'' + \frac{1}{x}\alpha' + \left(\xi_1^2 - \xi_2^2 + \frac{\lambda_2^2 - \lambda_1^2}{x^2}\right)\alpha - 2\left(\xi_2^2 - \frac{\lambda_2^2}{x^2}\right)\beta' - 2\frac{\lambda_2^2}{x^3}\beta,$$

also $R = 0$, $S = 0$. Hierin bedeuten, abweichend vom Wortlaut des Satzes 9, ξ_1, λ_1 und ξ_2, λ_2 die zu den Funktionen u_1 und u_2 gehörigen Werte von ξ, λ in der Differentialgleichung (64).

Man entwickle α und β nach fallenden Potenzen von x

$$\alpha = ax^r + \cdots, \quad \beta = bx^s + \cdots, \quad ab \neq 0;$$

dann gehen R und S über in

$$b(\xi_1^2 - \xi_2^2)x^s + 2arx^{r-1} + \cdots$$

und

$$a(\xi_1^2 - \xi_2^2)x^r - 2b\xi_2^2 s x^{s-1} + \cdots;$$

also ist $\xi_1^2 = \xi_2^2$. Ohne Beschränkung der Allgemeinheit sei dann $\xi_1^2 = 1$. Aus $R = 0$ und $S = 0$ ergibt sich durch Elimination von α die homogene lineare Differentialgleichung vierter Ordnung für β

$$x^4\beta'''' + 2x^3\beta''' + x^2(4x^2 - 2\lambda_1^2 - 2\lambda_2^2 + 1)\beta'' + x(8x^2 + 2\lambda_1^2 + 2\lambda_2^2 - 1)\beta'$$
$$+ (\lambda_1 + \lambda_2 + 1)(\lambda_1 + \lambda_2 - 1)(\lambda_1 - \lambda_2 + 1)(\lambda_1 - \lambda_2 - 1)\beta = 0.$$

Die algebraische Funktion β ist also höchstens bei 0 und ∞ singulär. Setzt man

$$\beta = x^r(a_0 + a_1 x + \cdots + a_l x^l),$$

wo $a_0 a_l \neq 0$ ist, so folgt durch Koeffizientenvergleich

$$4(r + l)(r + l + 1) = 0,$$

also die Ganzzahligkeit von r, und ferner

$$(r - \lambda_1 - \lambda_2 - 1)(r + \lambda_1 + \lambda_2 - 1)(r - \lambda_1 + \lambda_2 - 1)(r + \lambda_1 - \lambda_2 - 1) = 0.$$

Daher ist $\lambda_1 + \lambda_2$ oder $\lambda_1 - \lambda_2$ ganz rational, im Widerspruch zur Voraussetzung. Damit ist Satz 9 bewiesen.

Die Voraussetzungen des Satzes 9 sind, wie man leicht sieht, auch notwendig für seine Gültigkeit; z. B. gilt für ganz rationales $\lambda_1 - \lambda_2$ zwischen J_{λ_1}, J'_{λ_1}, J_{λ_2} und x eine algebraische Gleichung, wie aus (56) folgt.

Will man für 2λ auch ungerade Werte zulassen, so bedarf Satz 9 einer Ergänzung, über die noch folgendes bemerkt sei. In diesem Falle lassen sich J_λ, J'_λ, Y_λ algebraisch durch x und e^{ix} ausdrücken, und man hat zu untersuchen, ob ein Ausdruck

$$\Phi = \phi_0 + \phi_1 e^{\alpha_1 x} + \phi_2 e^{\alpha_2 x} + \cdots + \phi_n e^{\alpha_n x}$$

identisch in x verschwinden kann, falls $\alpha_1, \cdots, \alpha_n$ voneinander verschiedene Zahlen bedeuten und $\phi_0, \cdots, \phi_n$ Polynome in den Besselschen Funktionen des Satzes 9 und x sind. Es sei n möglichst klein. Ist $\Phi = 0$, so folgt

$$\frac{d\Phi}{dx} = \frac{d\phi_0}{dx} + \frac{d\phi_1}{dx}e^{\alpha_1 x} + \cdots + \frac{d\phi_n}{dx}e^{\alpha_n x} + \alpha_1\phi_1 e^{\alpha_1 x} + \cdots + \alpha_n\phi_n e^{\alpha_n x} = 0,$$

also müssen die beiden in $e^{\alpha_1 x}, \cdots, e^{\alpha_n x}$ linearen Funktionen Φ und $\frac{d\Phi}{dx}$ proportional sein,

$$\frac{d\log\phi_0}{dx} = \frac{d\log\phi_1}{dx} + \alpha_1 = \cdots = \frac{d\log\phi_n}{dx} + \alpha_n$$

(73) $$\phi_0 = c\phi_1 e^{\alpha_1 x}$$

also $n = 1$. Ersetzt man hierin x durch $2x$ und eliminiert $e^{\alpha_1 x}$ aus der so entstandenen Gleichung und (73), so erhält man einen Widerspruch zu Satz 9. Also ist Φ nicht identisch in x gleich 0.

VII.

Für Potenzreihen mit *endlichem* Konvergenzradius gelingt es nicht mehr, durch die früher entwickelte Methode Transzendenzsätze abzuleiten; doch kommt man in vielen Fällen zu Aussagen über Irrationalität.

Anstatt der E-Funktionen $\sum_{n=0}^{\infty} \frac{a_n}{b_n} \frac{x^n}{n!}$ sind Potenzreihen

$$y = \sum_{n=0}^{\infty} \frac{a_n}{b_n} x^n,$$

bei denen also der Faktor $n!$ im Nenner des allgemeinen Gliedes fehlt, zu betrachten, welche denselben Bedingungen 1., 2., 3. genügen, wie sie bei der Definition der E-Funktionen angegeben waren; nur soll die Bedingung über das Anwachsen der Konjugierten von a_n und des kleinsten gemeinschaftlichen Vielfachen von $b_1, \cdots, b_n$ durch die schärfere ersetzt werden, daß dieses Anwachsen nicht stärker geschieht als bei einer Potenz c^n mit geeigneter konstanter Basis c. Solche Funktionen mögen G-Funktionen genannt werden; zu ihnen gehört trivialerweise die geometrische Reihe. Ähnlich wie bei den E-Funktionen folgt, daß die G-Funktionen einen Ring bilden. Es ist trivial, daß die Ableitung einer G-Funktion wieder eine solche ist. Aus Satz 5 folgt ferner, daß auch das Integral $\int_0^x G(t)dt$ eine G-Funktion ist.

Nach einem Satze von Eisenstein ist jede bei $x = 0$ reguläre algebraische Funktion, die einer algebraischen Gleichung mit algebraischen Zahlenkoeffizienten genügt, eine G-Funktion. Das von 0 bis x erstreckte Integral einer solchen algebraischen Funktion ist daher ebenfalls eine G-Eunktion. Ein anderes Beispiel einer G-Funktion liefert die hypergeometrische Reihe

$$1 + \frac{\alpha\beta}{\gamma \cdot 1} x + \frac{\alpha(\alpha+1)\beta(\beta+1)}{\gamma(\gamma+1) \cdot 1 \cdot 2} x^2 + \cdots,$$

wenn α, β, γ rational sind.

Man kann diese Funktionen denselben Betrachtungen unterwerfen, die im vorhergehenden auf J_0 angewendet worden sind. Bei der Differentiation der Näherungsformen, wie sie im § 2 ausgeführt wurde, hat man im vorliegenden Fall zu beachten, daß die durch oftmalige Differentiation einer Näherungsform entstehenden Näherungsformen noch einen großen gemeinsamen Teiler in ihren Zahlenkoeffizienten haben; erst nach Division durch diesen Teiler wird eine erfolgreiche Restabschätzung möglich.

Bei der Untersuchung des Abelschen Integrals $\int_0^x y\,dt$ ist es notwendig, allgemeinere Näherungsformen einzuführen. Man hat nämlich als Koeffizienten der Näherungsformen nicht Polynome in den unabhängigen Variabeln x allein, sondern Polynome in x und y, oder ganze Funktionen aus dem durch y und x erzeugten algebraischen Funktionenkörper zuzulassen. Diese Verallgemeinerung bietet jedoch keine wesentliche Schwierigkeit.

Die Durchführung der Rechnung ergibt folgendes Resultat: Es sei y eine algebraische Funktion von x, in deren Gleichung die Koeffizienten algebraische Zahlen sind. Es sei $x = 0$ ein regulärer Punkt der Funktion. Das Abelsche Integral $\int_0^x y\,dt$ sei keine algebraische Funktion. Es genüge $\xi \neq 0$ einer algebraischen Gleichung l-ten Grades, deren Koeffizienten ganz rational und absolut $\leq M$ sind. Es sei ε irgendeine positive Zahl und

$$(74) \qquad |\xi| < c_7 e^{-(\log n)^{\frac{1}{2}+\varepsilon}},$$

wo c_l eine gewisse positive Funktion von l und ε bedeutet. Dann genügt die Zahl $\int_0^\xi y\,dt$ keiner algebraischen Gleichung l-ten Grades mit rationalen Koeffizienten.

Die Bedingung (74) ist zum Beispiel für $\xi = \frac{1}{n}$ und hinreichend großes natürliches n erfüllt, ebenfalls für $\xi = (\sqrt{2}-1)^n$ und hinreichend großes n, allgemeiner für jede hinreichend hohe Potenz jeder im Innern des Einheitskreises gelegenen algebraischen Zahl.

Für hinreichend großes natürliches n ist also etwa $\int_0^{\frac{1}{n}} \frac{dx}{\sqrt{1-x^4}}$ keine algebraische Irrationalität von kleinerem als 1000sten Grade; und allgemeiner kann man nach (74) unendlich viele algebraische ξ von vorgeschriebenem Grade angeben. so daß das elliptische Integral

$$\int_0^\xi \frac{dx}{\sqrt{4x^3-g_2x-g_3}}$$

mit algebraischen Werten von g_2 und g_3 keiner algebraischen Gleichung vorgeschriebenen Grades genügt.

Es sei $x=\eta$ eine algebraische Zahl und regulärer Punkt der algebraischen Funktion y. Falls $\int_0^\eta y\,dt = \gamma$ einer algebraischen Gleichung vom Grade $\leq l$ genügt, so bilde man

$$\int_0^x y(t)\,dt = \gamma + \int_0^{x-\eta} y(t+\eta)\,dt$$

und kann dann $x=\zeta$ in der Nähe von $x=\eta$ so wählen, daß $\zeta-\eta=\xi$ dem Körper von η angehört und der Ungleichung (74) genügt, daß ferner die Zahl $\int_0^\xi y(t+\eta)\,dt$ keiner Gleichung vom Grade $\leq l^2$ genügt. Dann genügt aber die Zahl $\int_0^\zeta y\,dt$ keiner Gleichung l-ten Grades. Damit ist bewiesen, daß für jeden imaginären Zahlkörper diejenigen Zahlen η des Körpers, für welche der Wert $\int_0^\eta y\,dt$ keine algebraische Irrationalität vom Grade $\leq l$ ist, überall dicht in der aufgeschnittenen x-Ebene liegen.

Die Abelschen Integrale, welche selbst algebraische Funktionen von x sind, haben offenbar nicht die soeben genannte Eigenschaft. Damit ist eine arithmetische Eigenschaft gefunden, welche die transzendenten Abelschen Integrale vor den algebraischen auszeichnet.

Wendet man diese Betrachtungen speziell auf die Funktion $\int_0^x \frac{dt}{1+t} = \log(1+x)$ an, so gelangt man infolge der Einschränkung (74) nicht bis zum Lindemannschen Satz. Es gelingt auf diesem Wege nicht, zu beweisen, daß die Zahlen $\log 2$, $\log 3$, $\cdots$ transzendent sind, es folgt nur, daß unter ihnen beliebig hohe Irrationalitäten auftreten, genauer, daß die Anzahl der Zahlen $\log 2, \cdots, \log n$, welche algebraische Irrationalitäten beschränkten Grades sind, für jedes $\varepsilon > 0$ sicherlich $o(n^\varepsilon)$ ist.

Auch auf die arithmetische Untersuchung von π und überhaupt der Perioden Abelscher Integrale lassen sich diese Überlegungen nicht anwenden; man kann der Bedingung (74) noch nicht einmal genügen, wenn man die Teilungstheorie zu Hilfe nimmt.

Man kann den Satz über Abelsche Integrale auch als eine Aussage über ihre Umkehrungsfunktion formulieren. So gilt z. B., daß die mit algebraischen g_2, g_3 gebildete Weierstrasssche $\wp$-Funktion für kein Argument, das algebraisch irrational vom l-ten Grade und hinreichend dicht am Nullpunkte gelegen ist, einen ganzen rationalen Wert annimmt.

Analoge Untersuchungen lassen sich für die hypergeometrische Reihe ausführen, wenn α, β, γ rationale Werte haben. Man hat dann zunächst die Sätze 1, 2, 3 zu übertragen und insbesondere alle hypergeometrischen Funktionen zu ermitteln, die algebraischen Differentialgleichungen erster Ordnung genügen. Dies geschieht nach der Methode des Beweises von Satz 2. Es stellt sich heraus, daß einzig und allein die Schwarzschen Ausnahmefälle in Betracht kommen.

Für die nichtalgebraischen hypergeometrischen Funktionen $F(\alpha, \beta, \gamma, x)$ mit rationalen α, β, γ kann man dann einen ganz entsprechenden Irrationalitätssatz aufstellen, wie er oben für die Abelschen Integrale ausgesprochen wurde; und durch die in diesem Satze formulierte arithmetische Eigenschaft werden dann auch wieder die nichtalgebraischen hypergeometrischen Reihen von den algebraischen getrennt.

Ein Beispiel liefert die spezielle Funktion

$$F\left(\frac{1}{2}, \frac{1}{2}, 1, x\right) = \sum_{n=0}^{\infty} \binom{2n}{n}^2 \left(\frac{x}{16}\right)^n = \frac{2}{\pi} \int_0^1 \frac{dt}{\sqrt{(1-t^2)(1-xt^2)}}$$

$$= \int_0^1 \frac{dt}{\sqrt{(1-t^2)(1-xt^2)}} : \int_0^1 \frac{dt}{\sqrt{1-t^2}}.$$

Man erhält das Resultat, daß ihr Wert für rationales $x = \frac{p}{q}$ irrational ist, falls

$$0 < \left|\frac{p}{q}\right| < c_8 10^{-\sqrt{\log|q|}}$$

gewählt wird, wo c_8 eine positive Konstante bedeutet; und ähnliche Sätze gelten für höhere Irrationalität. Für diese Werte des Moduls ist also die reelle Periode des elliptischen Integrals $\int \frac{dt}{\sqrt{(1-t^2)(1-xt^2)}}$ mit π inkommensurabel. Für die andere Periode bekommt man keine Aussage, da die zugehörige hypergeometrische Funktion im Nullpunkt logarithmisch verzweigt ist.

Es sei noch bemerkt, daß sich Zahlen wie $2^{\sqrt{2}}$ der Untersuchung entziehen, weil die Nenner der Koeffizienten in der binomischen Reihe $(1+x)^\lambda = 1 + \binom{\lambda}{1}x + \cdots$ zu stark anwachsen, wenn λ eine irrationale algebraische Zahl bedeutet.

VIII.

Auch die in VI. angestellten Untersuchungen lassen sich auf G-Funktionen übertragen. Man kommt dann zu Sätzen von der Art: Wenn algebraische Zahlen $\xi_1, \xi_2, \cdots$ der Bedingung (74) genügen, so besteht zwischen den Funktionswerten $G_1(\xi_1), G_2(\xi_2), \cdots$ keine algebraische Gleichung, deren Grad nicht allzu hoch ist. Auf diese Weise erhält man Aussagen über die Logarithmusfunktion, welche nicht im Lindemannschen Satz enthalten sind. Es ergibt sich so, daß in der Zahlenfolge

$$\frac{\log 2}{\log 3}, \frac{\log 3}{\log 4}, \frac{\log 4}{\log 5}, \ldots$$

beliebig hohe Irrationalitäten auftreten. Bisher war nur das triviale Resultat bekannt, daß diese Zahlen sämtlich irrational sind. Es gibt also unter den Briggischen Logarithmen der natürlichen Zahlen beliebig hohe Irrationalitäten, und dies gilt auch für jede Basis des Logarithmensystems. Allgemeiner lassen sich positive rationale Zahlen $r_1, \cdots, r_n$ auf unendlich viele Weisen so angeben, daß zwischen den Zahlen $\log r_1, \cdots, \log r_n$ keine algebraische Gleichung mit ganzen rationalen Koeffizienten besteht, deren Grad unterhalb einer vorgegebenen Zahl liegt, insbesondere also auch so, daß die n Logarithmen in einem vorgeschriebenen Körper linear unabhängig sind.

Endlich gewinnt man einen Zugang zu arithmetischen Sätzen über die Schwarzschen automorphen Funktionen, indem man die Gleichungen untersucht, welche zwischen den Werten verschiedener hypergeometrischer Funktionen mit algebraischem Argument bestehen können.

Zweiter Teil: Über diophantische Gleichungen.

ARTHUR SCHOENFLIES zum Gedächtnis.

Die mathematische Wissenschaft verdankt ANDRÉ WEIL den Beweis einer wichtigen arithmetischen Eigenschaft der algebraischen Kurven. Der Formulierung des WEILschen Satzes sei folgendes vorausgeschickt:

Es sei $f(x, y) = 0$ die Gleichung einer algebraischen Kurve vom Geschlecht $p > 0$. Die Koeffizienten des Polynoms f mögen einem algebraischen Zahlkörper $\mathfrak{K}$ angehören. Ein System von p Kurvenpunkten mit den Koordinaten $x_l, y_l (l = 1, \cdots, p)$ heiße *rational*, wenn alle in den p Paaren x_l, y_l symmetrischen rationalen Verbindungen mit Koeffizienten aus $\mathfrak{K}$ wieder in $\mathfrak{K}$ gelegen sind. Jedem System $\mathfrak{P}$ von p Curvenpunkten ordne man ein System $s_1, \cdots, s_p$ von p complexen Zahlen zu, indem man jedes der p zur Kurve gehörigen ABELschen Normalintegrale erster Gattung $w_k (k = 1, \cdots, p)$ von dem Punkte $x_l^{(0)}, y_l^{(0)}$ eines festen Punktsystems $\mathfrak{P}_0$ bis zu dem Punkte x_l, y_l von $\mathfrak{P}$ erstreckt und für jedes k der Reihe $1, \cdots, p$ die für $l = 1, \cdots, p$ entstehenden Werte addiert; die so entstehenden p Integralsummen seien $s_1, \cdots, s_p$. Es werde nun vorausgesetzt, daß das Punktsystem $\mathfrak{P}_0$ rational sei. Ist auch $\mathfrak{P}$ rational, so möge der Kürze wegen auch das System der p Zahlen $s_1, \cdots, s_p$ rational genannt werden. Nach dem ABELschen Theorem sind mit $s_1, \cdots, s_p$ und $t_1, \cdots, t_p$ auch die Systeme $s_1 + t_1, \cdots, s_p + t_p$ und $s_1 - t_1, \cdots, s_p - t_p$ rational. Die rationalen Systeme $s_1, \cdots, s_p$ bilden also einen Modul.

Der Satz von WEIL besagt, daß dieser Modul eine endliche Basis besitzt. Der Spezialfall $p = 1$ dieses Satzes wurde für den Körper der rationalen Zahlen bereits vor einigen Jahren von MORDELL entdeckt.

Durch den Satz von WEIL wird nahegelegt, das Theorem von FERMAT und allgemeiner die Theorie der algebraischen diophantischen Gleichungen mit zwei Unbekannten von einer neuen Seite anzugreifen. Doch dürfte wohl der Beweis der Vermutung, daß jede solche Gleichung, wenn ihr Geschlecht größer als 1 ist, nur endlich viele Lösungen in rationalen Zahlen besitzt, noch die Überwindung erheblicher Schwierigkeiten erfordern. Dagegen läßt sich, indem man WEILS Ideen mit den Überlegungen kombiniert, welche zum Beweis des Satzes von THUE und weiterhin im ersten Teile dieser Abhandlung zum Transzendenzbeweis der BESSELschen Funktion $J_0(\xi)$ für algebraisches $\xi \neq 0$ geführt haben, ein Resultat ableiten, das die Sonderstellung der linearen und der indefiniten quadratischen Gleichungen in der Theorie der algebraischen diophantischen Gleichungen mit zwei *ganzen* rationalen Unbekannten zum Ausdruck bringt.

Es sei $L(x, y) = ax + by$ eine homogene lineare Form mit ganzen rationalen Koeffizienten a und b. Ist c eine ganze rationale Zahl, in welcher der größte gemeinsame Teiler von a und b aufgeht, so hat nach BACHET die Gleichung $L(x, y) = c$ unendlich viele Lösungen in ganzen rationalen x, y.

Es sei $Q(x, y) = ax^2 + bxy + cy^2$ eine indefinite quadratische Form mit ganzen rationalen Koeffizienten a, b, c, deren Diskriminante $b^2 - 4ac$ keine Quadratzahl ist. Für alle durch Q darstellbaren ganzen rationalen Zahlen $d \neq 0$ hat nach LAGRANGE die Gleichung $Q(x, y) = d$ unendlich viele Lösungen in ganzen rationalen x, y.

Aus diesen beiden Typen von binären diophantischen Gleichungen mit unendlich vielen ganzzahligen Lösungen erhält man leicht allgemeinere. Es seien nämlich $A(u, v)$ und $B(u, v)$ irgend zwei homogene Polynome n-ten Grades, die ganze rationale Koeffizienten haben und nicht beide zu L^n proportional sind. Durch Elimination von $u : v$ aus den Gleichungen

$$x = \frac{A(u, v)}{L^n(u, v)}, \qquad y = \frac{B(u, v)}{L^n(u, v)}$$

erhält man dann eine algebraische Gleichung zwischen x und y. Diese Gleichung hat sicherlich unendlich viele Lösungen in ganzen rationalen x, y, wenn noch vorausgesetzt wird, daß der größte gemeinsame Teiler der Koeffizienten a und b von L zur n-ten Potenz in den Koeffizienten von A und B aufgeht. Sind ferner $C(u, v)$ und $D(u, v)$ zwei ganzzahlige homogene Polynome vom Grade $2n$, die nicht beide zu Q^n proportional sind, so wird durch

$$x = \frac{C(u, v)}{Q^n(u, v)}, \qquad y = \frac{D(u, v)}{Q^n(u, v)}$$

ebenfalls eine algebraische Gleichung zwischen x und y definiert, welche unendlich viele ganzzahlige Lösungen besitzt, wenn noch vorausgesetzt wird, daß es eine durch Q darstellbare Zahl gibt, deren n-te Potenz in allen Koeffizienten von C und D enthalten ist.

Die beiden genannten Typen von diophantischen Gleichungen mit unendlich vielen ganzzahligen Lösungen sind auf einfache Weise aus den klassischen Gleichungen $L = c$ und $Q = d$ abgeleitet. In dieser Abhandlung soll nun bewiesen werden, daß hierdurch bereits *alle* binären diophantischen Gleichungen mit unendlich vielen Lösungen in ganzen rationalen Zahlen umfaßt werden. Es wird also behauptet:

I. *Die algebraische Gleichung $f(x, y) = 0$ sei nicht dadurch identisch in einem Parameter t lösbar, daß man entweder $x = A : L^n$, $y = B : L^n$ oder $x = C : Q^n$, $y = D : Q^n$ setzt, wo A, B, C, D ganzzahlige Polynome in t, L ein lineares, Q ein indefinites quadratisches Polynom in t bedeuten. Dann hat sie nur endlich viele Lösungen in ganzen rationalen Zahlen.*

Der Ausnahmefall erfordert offenbar, daß das Geschlecht der Gleichung den Wert $p = 0$ habe. Es wird zunächst nach Maillet durch eine einfache Betrachtung gezeigt werden, daß die Behauptung für $p = 0$ richtig ist. Die eigentliche Schwierigkeit bietet der Fall $p > 0$, also der Beweis des Satzes:

Eine algebraische Gleichung, deren Geschlecht positiv ist, hat nur endlich viele Lösungen in ganzen rationalen Zahlen.

Es macht die gleiche Mühe, den analogen Satz für algebraische Zahlkörper zu beweisen. Dabei ist es für die Formulierung des Satzes zweckmäßig, die Voraussetzungen noch etwas zu erweitern. Bei der Lösung der Gleichung $f(x, y) = 0$ sollen nunmehr für x und y auch gebrochene Zahlen des algebraischen Zahlkörpers $\mathfrak{K}$ zugelassen werden, aber nur solche gebrochenen Zahlen, von denen ein Multiplum cx, cy, wo c eine *feste* natürliche Zahl bedeutet, wieder ganz ist. Solche Zahlen mögen kurz *ganzartig* genannt werden. Und nun gilt folgende Verallgemeinerung des oben ausgesprochenen Satzes:

II. *Damit die irreduzible Gleichung $f(x, y) = 0$ in irgendeinem algebraischen Zahlkörper unendlich viele Lösungen in ganzartigen Zahlen besitzt, ist notwendig und hinreichend, daß der Gleichung $f = 0$ durch zwei rationale Funktionen $x = P(t)$, $y = Q(t)$ identisch in t genügt werden kann; dabei sollen P und Q die Gestalt*

$$P(t) = a_n t^n + a_{n-1} t^{n-1} + \cdots + a_{-n} t^{-n} \tag{75}$$

$$Q(t) = b_n t^n + b_{n-1} t^{n-1} + \cdots + b_{-n} t^{-n} \tag{76}$$

besitzen und nicht beide konstant sein.

Dies läßt sich auch noch anders ausdrücken. Man kann offenbar voraussetzen, daß t eine rationale Funktion von x und y mit algebraischen Koeffizienten ist. Zunächst folgt nämlich aus der Darstellung $x = P$, $y = Q$, daß f vom Geschlecht 0 ist. Nun drücke man x und y aus als rationale Funktionen eines Parameters τ, der seinerseits wieder rationale Funktion von x und y ist; dann ist τ eine rationale Funktion von t. Da τ nur bis auf eine lineare Transformation bestimmt ist, so darf man annehmen, daß für $t = 0$ auch $\tau = 0$ und für $t = \infty$ entweder $\tau = \infty$ oder $\tau = 0$ ist. Es haben aber P und Q als Funktionen von t keine anderen Pole als höchstens die Stellen $0, \infty$, also gilt dasselbe für die Abhängigkeit von τ. Es gelten also auch Gleichungen der Form (75) und (76) mit τ statt t, und man kann annehmen, daß t bereits der uniformisierende Parameter ist. Die Koeffizienten $a_n, \cdots, b_{-n}$ in (75) und (76) bestimmen sich durch Einsetzen von $x = P(t)$, $y = Q(t)$ in $f = 0$; sie können daher als algebraische Zahlen gewählt werden. Dasselbe gilt für die Koeffizienten der rationalen Funktion $R(x, y)$, welche t durch x und y ausdrückt. Ist nun Q nicht konstant, so geht durch die Substitution

$$x = P(t) + u\,, \quad y = Q(t)$$

die Gleichung $f(x, y) = 0$ in $u = 0$ über. Ist aber Q konstant, also P nicht konstant, so erhält man das gleiche Resultat durch die Substitution

$$x = P(t) + u\,, \quad y = Q(t) - u\,.$$

Enthalten nun P und Q keine negativen Potenzen von t, so werden durch die auf der Kurve birationale Substitution

$$x = P(t) + u\,, \quad y = Q(t)\,; \qquad u = 0\,, \quad t = R(x, y)$$

bzw. $$x = P(t) + u\,, \quad y = Q(t) - u\,; \quad u = 0\,, \quad t = R(x, y)$$

alle Punkte der Kurve $f = 0$, deren Koordinaten x, y ganzartige Zahlen eines algebraischen Zahlkörpers sind, übergeführt in Punkte der Kurve $u = 0$, deren Koordinaten u, t ebenfalls ganzartige Zahlen eines algebraischen Zahlkörpers sind; und umgekehrt. Enthalten P und Q keine positiven Potenzen von t, so ersetze man t durch $\frac{1}{t}$ und kommt auf den soeben behandelten Fall zurück. Kommen endlich in einer der Funktionen P, Q positive und negative Potenzen von t vor, so ersetze man in den negativen Potenzen $\frac{1}{t}$ durch u. Dann sind durch die birationale Substitution

$$x = a_n t^n + \cdots + a_1 t + a_0 + a_{-1} u + \cdots + a_{-n} u^n, \quad y = b_n t^n + \cdots + b_1 t + b_0 + b_{-1} u + \cdots + b_{-n} u^n$$
$$u = 1 : R(x, y), \quad t = R(x, y)$$

wieder die Punkte von $f = 0$, deren Koordinaten x, y ganzartige Zahlen eines algebraischen Zahlkörpers sind, übergeführt in ebensolche Punkte u, t der Kurve $ut = 1$; und umgekehrt. Daher läßt sich der Satz folgendermaßen fassen:

Damit $f(x, y) = 0$ in einem algebraischen Zahlkörper unendlich viele ganzartige Lösungen besitzt, ist notwendig und hinreichend, daß sich die Gleichung $f = 0$ entweder in $u = 0$ oder in $ut = 1$ überführen läßt, und zwar durch eine birationale Transformation, welche alle ganzartigen Paare x, y und u, t miteinander verknüpft.

Sieht man von solchen birationalen Transformationen ab, so haben also nur die trivialen Gleichungen $u = 0$ und $ut = 1$ unendlich viele Lösungen in ganzartigen Zahlen eines algebraischen Zahlkörpers.

Beachtet man, daß t auf der RIEMANNschen Fläche des durch $f = 0$ definierten algebraischen Gebildes jeden Wert genau einmal annimmt, so erkennt man, daß sich die beiden Fälle $u = 0$ und $ut = 1$ folgendermaßen trennen lassen: Im ersten Falle wird $|x| + |y|$ auf der RIEMANNschen Fläche in genau einem Punkte ∞, im zweiten Falle in genau zwei Punkten; und dies ist auch wiederum hinreichend für das Bestehen von (75) und (76). Also hat man den sehr einfachen Satz:

Damit die Gleichung $f(x, y) = 0$ unendlich viele ganzartige Lösungen in einem algebraischen Zahlkörper besitzt, ist notwendig und hinreichend, daß die zugehörige RIEMANNsche Fläche vom Geschlecht 0 ist und höchstens zwei Unendlichkeitsstellen der Funktion $|x| + |y|$ enthält.

Die beiden Typen $u = 0$ und $ut = 1$ sind wesentlich voneinander verschieden. Das äußert sich auch in der Anzahl der Lösungen, die unterhalb einer festen Schranke liegen. Betrachtet man nämlich nur diejenigen ganzartigen Lösungen x, y aus einem algebraischen Zahlkörper, für welche die absoluten Beträge aller Konjugierten von x und y unterhalb der Schranke M liegen, so ist, wie sich aus dem Vorhergehenden unschwer ergibt, im ersten Fall die genaue Größenordnung der Lösungsanzahl gleich $M^\varkappa$, im zweiten Fall gleich $(\log M)^\lambda$, wo $\varkappa$ und λ gewisse positive Zahlen bedeuten, die nur von f und dem Körper abhängen.

Beim Beweise ergibt sich implizite auch eine endliche obere Schranke für die Anzahl der Lösungen jeder Gleichung $f = 0$, die nicht einem der beiden Ausnahmetypen angehört. Eine Schranke für die Lösungen selbst ergibt sich aber dabei nicht; das Problem der Aufsuchung der endlich vielen Lösungen bleibt also unerledigt.

Die Methode des Beweises liefert auch ohne weiteres ein analoges Resultat für Raumkurven: Ein System von $n - 1$ unabhängigen algebraischen Gleichungen in n Unbekannten $x_1, \cdots, x_n$ mit algebraischen Koeffizienten besitzt dann und nur dann unendlich viele ganzartige Lösungen in einem algebraischen Zahlkörper, wenn für $x_1, \cdots, x_n$ eine Parameterdarstellung durch Polynome in t und t^{-1} besteht.

Eine andere Verallgemeinerung bezieht sich auf den Fall $p > 0$. Es wird nämlich die Endlichkeit der Lösungsanzahl von $f(x, y) = 0$ bereits unter der geringeren Voraussetzung bewiesen werden, daß nur eine der Unbekannten ganzartig ist. Hat also $f = 0$ unendlich viele Lösungen x, y in einem algebraischen Zahlkörper, so ist für diese weder die Norm des Nenners von x noch die Norm des Nenners von y beschränkt.

Endlich sei noch eine Anwendung auf den HILBERTschen Irreduzibilitätssatz erwähnt. Es bedeute $P(w, x, y, \cdots)$ ein Polynom der Variabeln $w, x, y, \cdots$, das ganze Koeffizienten aus einem algebraischen Zahlkörper $\mathfrak{K}$ besitzt. Faßt man P als Funktion von $x, y, \cdots$, allein auf, so sind die Koeffizienten Polynome in w; sie seien in irgendeiner Reihenfolge mit $a_1(w), a_2(w), \cdots$ bezeichnet. Für einen bestimmten Zahlenwert von w sei nun P in $\mathfrak{K}$ reduzibel. Dann ist $P = QR$, wo Q und R Polynome in $x, y, \cdots$ bedeuten, deren Koeffizienten $b_1, b_2, \cdots$ und $c_1, c_2, \cdots$ in $\mathfrak{K}$ liegen. Wird von einem konstanten Faktor abgesehen, so bestehen bei festem w für Q und R nur endlich viele Möglichkeiten, die Produkte $b_k c_l = d_{kl}$ $(k = 1, 2, \cdots; l = 1, 2, \cdots)$ sind also durch w endlich vieldeutig bestimmt. Nach einem Satze von GAUSS und KRONECKER sind diese Produkte ganz, wenn w es ist. Die Matrix (d_{kl}) hat den Rang 1. Durch Koeffizientenvergleich ergeben sich ferner die Zahlen $a_1(w), a_2(w), \cdots$ als Summen gewisser der d_{kl}. Dies liefert ein System von algebraischen Gleichungen in w und den d_{kl}. Hierdurch wird für beliebiges w im Raume der w und d_{kl} entweder ein System von endlich vielen Punkten oder aber eine Raumkurve bestimmt. Die notwendige und hinreichende Bedingung dafür, daß auf dieser Kurve unendlich viele Punkte liegen, deren Koordinaten w und d_{kl} $(k = 1, 2, \cdots; l = 1, 2, \cdots)$ ganzartige Zahlen eines algebraischen Zahlkörpers sind, ist nun oben angegeben worden. Es folgt:

Für die Existenz eines algebraischen Zahlkörpers $\mathfrak{K}$, in welchem das Polynom P für unendlich viele ganzartige Werte aus $\mathfrak{K}$ reduzibel ist, ist notwendig und hinreichend, daß P nach einer geeigneten Substitution der Form

$$w = \alpha_n t^n + \alpha_{n-1} t^{n-1} + \cdots + \alpha_{-n} t^{-n}$$

identisch in t reduzibel ist.

Nach Früherem gewinnt man in diesem Ausnahmefall auch eine genaue Abschätzung der Dichtigkeit derjenigen unendlich vielen ganzartigen w, die P reduzibel machen.

§ 1. Gleichungen vom Geschlecht 0.

Ist die Gleichung $f(x, y) = 0$ vom Geschlecht 0, so gilt die Uniformisierung

$$(77)\qquad x = \frac{\phi(u, v)}{\chi(u, v)}, \quad y = \frac{\psi(u, v)}{\chi(u, v)}, \quad \frac{u}{v} = \frac{A(x, y)}{B(x, y)};$$

dabei bedeuten ϕ, ψ, χ homogene Polynome gleichen Grades und A, B Polynome. Die Koeffizienten von f mögen dem algebraischen Zahlkörper $\mathfrak{K}$ angehören. Es werde angenommen, daß es einen regulären Punkt der Kurve $f = 0$ gibt, dessen Koordinaten $x = \xi$, $y = \eta$ in $\mathfrak{K}$ liegen. Ist in diesem Punkte $\frac{dy}{dx} = \infty$, so vertausche man die Bedeutung von y und x.

Durch (77) ist die Funktion $\frac{u}{v} = t$ auf der Kurve bis auf eine lineare Transformation bestimmt. Da t auf der Kurve jeden Wert genau einmal annimmt, so ist insbesondere in dem regulären Punkte ξ, η der Wert $\frac{dt}{dx}$ von 0 verschieden. Man kann also voraussetzen, daß in diesem Punkte t den Wert 0 hat und $\frac{dt}{dx}$ den Wert 1. Endlich darf man noch annehmen, daß im Punkt ξ, η der Wert $\frac{d^2 t}{dx^2} = 0$ ist, denn $\frac{t}{1 + ct}$ hat dort die zweite Ableitung $\frac{d^2 t}{dx^2} - 2c$. Durch die genannten drei Eigenschaften ist dann t vollständig bestimmt. Es soll nun gezeigt werden, daß die Koeffizienten der Polynome ϕ, ψ, χ, A, B als ganze Zahlen aus $\mathfrak{K}$ gewählt werden können. Jedenfalls können diese Koeffizienten als algebraische Zahlen gewählt werden; diese mögen einem Körper $\mathfrak{L}$ angehören, der $\mathfrak{K}$ enthält. Man ersetze nun in (77) alle Koeffizienten durch die Werte, die ihnen in irgendeinem zu $\mathfrak{L}$ in bezug auf $\mathfrak{K}$ konjugierten Körper $\mathfrak{L}'$ entsprechen. Da man wieder eine Uniformisierung erhält, so geht die neue Funktion t aus der alten durch lineare Substitution hervor, und da die Werte 0, 1, 0 beim Übergang von $\mathfrak{L}$ zu $\mathfrak{L}'$ invariant bleiben, so gilt das gleiche von der Funktion t. Bildet man nun in den drei

Gleichungen (77) das arithmetische Mittel aus allen in bezug auf $\mathfrak{K}$ konjugierten Gleichungen, so werden alle Koeffizienten der rechten Seiten Zahlen aus $\mathfrak{K}$.

Man darf annehmen, daß die drei Polynome ϕ, ψ, χ teilerfremd sind. Auf Grund des euklidischen Algorithmus gilt dann

$$P_1\phi + Q_1\psi + R_1\chi = \lambda_1 u^h \tag{78}$$

$$P_2\phi + Q_2\psi + R_2\chi = \lambda_2 v^h; \tag{79}$$

dabei bedeuten P_1, Q_1, R_1, P_2, Q_2, R_2 homogene Polynome in u, v mit ganzen Koeffizienten aus $\mathfrak{K}$, h eine natürliche Zahl und λ_1, λ_2 ganze von o verschiedene Zahlen aus $\mathfrak{K}$. Sind x und y Zahlen aus $\mathfrak{K}$, so sind nach (77) die Werte u und v als ganze Zahlen aus $\mathfrak{K}$ wählbar. Es sei $(u, v) = \delta$ ein größter gemeinsamer Teiler von u und v. Sind nun x und y ganze Zahlen aus $\mathfrak{K}$, so geht nach (77) die Zahl $\chi\left(\frac{u}{\delta}, \frac{v}{\delta}\right)$ in den Zahlen $\phi\left(\frac{u}{\delta}, \frac{v}{\delta}\right)$ und $\psi\left(\frac{u}{\delta}, \frac{v}{\delta}\right)$ auf, also nach (78) und (79) auch in den Zahlen $\lambda_1\left(\frac{u}{\delta}\right)^h$ und $\lambda_2\left(\frac{v}{\delta}\right)^h$, also auch in $\lambda_1\lambda_2$. Man hat daher eine Gleichung

$$\chi\left(\frac{u}{\delta}, \frac{v}{\delta}\right) = \gamma, \tag{80}$$

wo γ ein Teiler der festen Zahl $\lambda_1\lambda_2$ ist.

Zunächst sei nun $\mathfrak{K}$ der Körper der rationalen Zahlen. Dann ist γ eine von endlich vielen ganzen rationalen Zahlen; χ ist ein homogenes Polynom mit ganzen rationalen Koeffizienten; $\frac{u}{\delta}$ und $\frac{v}{\delta}$ sind ganze rationale Zahlen. Nach dem Thueschen Satze kann die Gleichung (80) nur dann unendlich viele Lösungen $\frac{u}{\delta}$, $\frac{v}{\delta}$ haben, wenn $\chi(u, v)$ Potenz eines linearen oder eines indefiniten quadratischen Polynoms ist. Andererseits gehören zu verschiedenen Paaren x, y auch verschiedene Werte $\frac{u}{v}$. Damit ist die Behauptung I. im Falle $p = 0$ bewiesen.

Nunmehr sei $\mathfrak{K}$ ein beliebiger algebraischer Zahlkörper. Sind von den linearen Faktoren von $\chi(u, v)$ nur höchstens zwei voneinander verschieden, werden also x und y nur für höchstens zwei Werte von $u:v$ unendlich groß, so schaffe man durch eine lineare Transformation von $u:v$ diese Werte nach o und ∞ und erhält dann für x und y Ausdrücke der Form (75) und (76). Enthält $\chi(u, v)$ mehr als zwei verschiedene Linearfaktoren $u-\alpha_1 v$, $u-\alpha_2 v$, $u-\alpha_3 v$, $\cdots$, so kann man ohne Beschränkung der Allgemeinheit α_1, α_2, α_3, $\cdots$ als ganze Zahlen aus $\mathfrak{K}$ voraussetzen, da man ja sonst nur v gleich einem geeigneten Vielfachen einer neuen Variabeln zu wählen und zu einem Oberkörper von $\mathfrak{K}$ überzugehen hätte. Dann sind aber die Zahlen

$$\frac{u}{\delta}-\alpha_1\frac{v}{\delta}, \quad \frac{u}{\delta}-\alpha_2\frac{v}{\delta}, \quad \frac{u}{\delta}-\alpha_3\frac{v}{\delta}$$

sämtlich Divisoren der festen Zahl $\lambda_1\lambda_2$. In der Gleichung

$$\frac{\alpha_3-\alpha_2}{\alpha_1-\alpha_2}\,\frac{u-\alpha_1 v}{u-\alpha_3 v} + \frac{\alpha_3-\alpha_1}{\alpha_2-\alpha_1}\,\frac{u-\alpha_2 v}{u-\alpha_3 v} = 1 \tag{81}$$

kommen also für $\frac{u-\alpha_1 v}{u-\alpha_3 v}$ und $\frac{u-\alpha_2 v}{u-\alpha_3 v}$ nur endlich viele nicht assoziierte Zahlen in Betracht. Es sei n eine natürliche Zahl. Nach Dirichlet ist die Gruppe der n-ten Potenzen der Einheiten von $\mathfrak{K}$ in bezug auf die Gruppe aller Einheiten von $\mathfrak{K}$ von endlichem Index. Daher gilt

$$(82) \qquad \frac{u-\alpha_1 v}{u-\alpha_3 v} = \gamma_1 \varepsilon_1^n, \quad \frac{u-\alpha_2 v}{u-\alpha_3 v} = \gamma_2 \varepsilon_2^n,$$

wo für die Zahlen γ_1 und γ_2 nur endlich viele Werte in Betracht kommen und ε_1, ε_2 Einheiten, also ganze Zahlen aus $\mathfrak{K}$ sind. Für jedes der endlich vielen Wertepaare γ_1, γ_2 hat man nach (81)

$$(83) \qquad \frac{\alpha_3-\alpha_2}{\alpha_1-\alpha_2}\gamma_1\varepsilon_1^n + \frac{\alpha_3-\alpha_1}{\alpha_2-\alpha_1}\gamma_2\varepsilon_2^n = 1 .$$

Nach einer Verallgemeinerung des Thueschen Satzes, die weiter unten in noch allgemeinerem Rahmen bewiesen werden wird, aber bereits früher bekannt war, hat die Gleichung (83) nur endlich viele Lösungen in ganzen Zahlen ε_1, ε_2 aus $\mathfrak{K}$, sobald der Grad n eine nur vom Körper abhängige Schranke übersteigt. Man hat also nach (82) auch nur endlich viele Werte $\frac{u}{v}$ und nach (77) nur endlich viele ganzzahlige in $\mathfrak{K}$ gelegene Lösungen x, y der Gleichung $f(x, y) = 0$.

Damit ist die Notwendigkeit der in der Behauptung II. ausgesprochenen Bedingung für die Existenz unendlich vieler ganzzahliger Lösungen aus $\mathfrak{K}$ im Falle $p = 0$ bewiesen. Den Fall der Ganzartigkeit führt man natürlich sofort auf den Fall der Ganzzahligkeit zurück, indem man x, y durch $\frac{x}{c}, \frac{y}{c}$ ersetzt, mit geeignetem festen natürlichen c. Daß die Bedingung der Behauptung II. hinreichend ist für die Lösbarkeit von $f = 0$ in unendlich vielen ganzartigen Zahlen eines geeigneten algebraischen Zahlkörpers, ist trivial; man braucht ja z. B. in (75) und (76) nur t durch alle Potenzen von $1+\sqrt{2}$ zu ersetzen und erhält unendlich viele ganzartige Zahlen x, y, die sämtlich einem algebraischen Zahlkörper angehören. Die Behauptung II. ist daher für den Fall $p = 0$ bewiesen.

Der Beweis gelang infolge der Zurückführung der gegebenen Gleichung auf die den Thueschen Methoden zugängliche Gleichung (83). Auch im Falle $p > 0$ soll nun die in ganzen Zahlen zu lösende Gleichung $f = 0$ transformiert werden in eine andere diophantische Gleichung, deren ganzzahlige Lösungen eine starke Annäherung an eine feste algebraische Zahl liefern; und die in der Einleitung dieser Abhandlung angedeuteten Überlegungen werden dann zeigen, daß eine solche starke Annäherung nur eine endliche Anzahl von Malen möglich ist. Die Analogie zwischen algebraischer und arithmetischer Teilbarkeit, die in (78) und (79) zum Ausdruck kommt und in obigem Beweise verwendet wurde, überträgt sich auf algebraische Funktionen, wie von Weil bemerkt wurde; der von ihm entdeckte Parallelismus zwischen Funktionenidealen und Zahlenidealen wird in § 2 mit geringfügigen Änderungen dargestellt.

§ 2. Funktionenideale und Zahlenideale.

Die Variabeln x und y seien durch eine Gleichung $f(x, y) = 0$ vom Geschlecht p verbunden. Man ersetze x, y durch $\frac{x}{z}, \frac{y}{z}$ und betrachte den Ring $\mathfrak{R}$ derjenigen homogenen rationalen Funktionen von x, y, z, welche für kein endliches Wertsystem x, y, z unendlich werden und als Koeffizienten algebraische Zahlen haben. Man setze noch fest, daß die Variabeln x, y, z nur teilerfremde ganzzahlige algebraische Werte annehmen sollen; dann sind auch die Werte der Funktionen von $\mathfrak{R}$ algebraische Zahlen.

Nun seien ϕ, ψ, χ irgend drei Formen von $\mathfrak{R}$, die denselben Grad n haben, und es sei der Quotient $\psi : \chi$ nicht konstant. Dann ist $\phi : \chi$ eine algebraische Funktion von $\psi : \chi$, und es gilt eine irreduzible Gleichung

$$(84) \qquad \phi^m + R_1(\psi, \chi)\phi^{m-1} + \cdots + R_m(\psi, \chi) = 0,$$

deren Koeffizienten $R_1(\psi, \chi), \cdots, R_m(\psi, \chi)$ homogene rationale Funktionen von ψ, χ mit den Graden $1, \cdots, m$ und algebraischen Zahlenkoeffizienten bedeuten. Es soll nun gezeigt werden, daß $R_1, \cdots, R_m$ Polynome sind, falls in keinem Kurvenpunkte die beiden

Formen ψ und χ zugleich von höherer Ordnung verschwinden als die Form ϕ. Hätte nämlich eine der Funktionen $R_1, \cdots, R_m$ einen linearen Faktor $a\psi + b\chi$ im Nenner, so würde bei *beliebigen* konstanten α, β die Funktion $\phi : (\alpha\psi + \beta\chi)$ unendlich für $a\psi + b\chi = 0$; also wäre dann zugleich $\psi = 0$, $\chi = 0$; setzt man nun $\alpha = 0$ oder $\beta = 0$, so folgt ein Widerspruch gegen die Voraussetzung betreffs des Verschwindens von ϕ. Nun sei c eine solche ganze Zahl $\neq 0$, daß die Koeffizienten der Polynome $cR_1, \cdots, cR_m$ ganz sind. Aus (84) folgt jetzt:

Wenn die beiden Formen ψ und χ in keinem Punkte zugleich von höherer Ordnung 0 werden als die Form ϕ, so existiert eine Konstante $c \neq 0$ derart, daß jeder gemeinsame Zahlenteiler von ψ und χ in $c\phi$ aufgeht.

Wählt man speziell $\psi = (a_1 x + b_1 y + c_1 z)^n$, $\chi = (a_2 x + b_2 y + c_2 z)^n$, wo die Koeffizienten $a_1, \cdots, c_2$ derartige ganze Zahlen sind, daß ψ und χ nicht simultan verschwinden, so folgt für jede Form ϕ von $\mathfrak{R}$ die Existenz einer Konstanten $c \neq 0$, so daß $c\phi$ eine ganze Zahl ist.

Allgemeiner seien $\phi_1, \cdots, \phi_k$ irgendwelche Formen von $\mathfrak{R}$, die nicht denselben Grad zu besitzen brauchen, und Φ sei eine solche Form, daß in keinem Punkte die Funktionen $\phi_1, \cdots, \phi_k$ simultan von höherer Ordnung verschwinden als Φ selber. Sind $\phi_1, \cdots, \phi_k$ sämtlich proportional, so sind $\Phi : \phi_1, \cdots, \Phi : \phi_k$ wieder Funktionen von $\mathfrak{R}$; also existiert eine Konstante $c \neq 0$, so daß $c\Phi : \phi_1, \cdots, c\Phi : \phi_k$ ganze Zahlen sind, und jeder gemeinsame Zahlenteiler von $\phi_1, \cdots, \phi_k$ geht auch in $c\Phi$ auf. Sind aber $\phi_1, \cdots, \phi_k$ nicht sämtlich proportional, so bestimme man die Formen $a_1, \cdots, a_k$ und $b_1, \cdots, b_k$ derart, daß

$$a_1\phi_1 + \cdots + a_k\phi_k = \psi, \quad b_1\phi_1 + \cdots + b_k\phi_k = \chi$$

beide homogen sind, und zwar von einem Grade, der um eine nicht negative Zahl r größer ist als der Grad von Φ; ferner sollen ψ und χ ebenfalls an keiner Stelle simultan von höherer Ordnung 0 werden als Φ. Versteht man nun unter ϕ nacheinander die 3 Funktionen $x^r\Phi$, $y^r\Phi$, $z^r\Phi$, so folgt aus dem oben Bewiesenen die Existenz einer Konstanten $c \neq 0$, für welche jeder gemeinsame Zahlenteiler von $\phi_1, \cdots, \phi_k$ in $c\Phi$ aufgeht. Haben nun die Formen $\phi_1, \cdots, \phi_k$ genau dieselben Nullstellen gemeinsam wie die Formen $\psi_1, \cdots, \psi_l$, und zwar auch mit der gleichen Vielfachheit, so unterscheiden sich also die größten gemeinsamen Teiler der Zahlen $\phi_1, \cdots, \phi_k$ von den größten gemeinsamen Teilern der Zahlen $\psi_1, \cdots, \psi_l$ nur um einen Faktor, dessen ganzzahliger Zähler und Nenner in einer festen vom Kurvenpunkte unabhängigen Zahl aufgeht. Ein solcher Faktor möge kurz *unitär* genannt werden. Der größte gemeinsame Teiler der Zahlen $\phi_1, \cdots, \phi_k$ ist daher bis auf einen unitären Faktor bereits durch die gemeinsamen Nullstellen der Funktionen $\phi_1, \cdots, \phi_k$ bestimmt.

Der Kurvenpunkt mit den homogenen Koordinaten x, y, z möge $\mathfrak{p}$ genannt werden. Ist $\mathfrak{p} = \mathfrak{p}_0$ ein fester Punkt, so konstruiere man zwei Funktionen ϕ_1 und ϕ_2, die nur die Nullstelle $\mathfrak{p}_0$ gemeinsam haben. Der größte gemeinsame Teiler der Zahlen ϕ_1 und ϕ_2 ist dann eine Funktion von $\mathfrak{p}$, die mit $\omega(\mathfrak{p}, \mathfrak{p}_0)$ bezeichnet werde; sie hängt allerdings noch von der Wahl der Formen ϕ_1, ϕ_2 ab, aber bei festem $\mathfrak{p}_0$ nur bis auf einen unitären Faktor. Man beachte nun, daß man aus den kl Produkten $\phi_1\psi_1, \cdots, \phi_k\psi_l$ durch lineare Kombination zwei Formen bilden kann, deren gemeinsame Nullstellen genau von den gemeinsamen Nullstellen von $\phi_1, \cdots, \phi_k$ und den gemeinsamen Nullstellen von $\psi_1, \cdots, \psi_l$ gebildet werden. Daher ist der größte gemeinsame Teiler der Zahlen $\phi, \psi, \cdots$, falls die Formen $\phi, \psi, \cdots$ die Nullstellen $\mathfrak{p}_1, \cdots, \mathfrak{p}_r$ gemeinsam haben, bis auf einen unitären Faktor gleich $\omega(\mathfrak{p}, \mathfrak{p}_1) \cdots \omega(\mathfrak{p}, \mathfrak{p}_r)$.

Für das Folgende benötigt man eine Abschätzung der arithmetischen Norm von $\omega(\mathfrak{p}, \mathfrak{p}_0)$ als Funktion von $\mathfrak{p}$. Die Ordnung der Kurve $f = 0$ sei h. Die Form

$$L = ax + by + cz$$

verschwindet in h Punkten. Nach dem Abelschen Theorem gibt es eine Funktion des durch $\frac{x}{z}$ und $\frac{y}{z}$ erzeugten algebraischen Funktionenkörpers, welche keine andern Pole hat als höchstens die hm Nullstellen von L^m und außerdem in $hm - p$ vorgeschriebenen Punkten verschwindet; diese Funktion heiße ϕ. Dann ist $\phi L^m = \psi$ eine Form aus $\mathfrak{R}$, deren Grad gleich m ist und die in $hm - p$ vorgeschriebenen Punkten verschwindet. Als diese Punkte wähle man den $(hm-p)$-mal gezählten Punkt $\mathfrak{p}_0$. Die weiteren p Nullstellen von ψ seien $\mathfrak{q}_1, \cdots, \mathfrak{q}_p$; sie hängen von $\mathfrak{p}_0$ und den Koeffizienten von L ab. Die Koeffizienten von ψ sind algebraische Zahlen, und bis auf einen unitären Faktor ist ψ gleich $(\omega(\mathfrak{p}, \mathfrak{p}_0))^{hm-p}\omega(\mathfrak{p}, \mathfrak{q}_1)\cdots\omega(\mathfrak{p}, \mathfrak{q}_p)$. Andererseits ist ψ für alle x, y, z endlich, also für $|x|+|y|+|z|=1$ beschränkt; und aus der Homogenität folgt nunmehr die Abschätzung

$$(85) \qquad |\psi| < c(|x|+|y|+|z|)^m,$$

wo c von $\mathfrak{p}_0$ und m, aber nicht von x, y, z abhängt.

Es mögen nun die Koordinaten x, y, z von $\mathfrak{p}$ in dem Körper $\mathfrak{K}$ der Koeffizienten von f gelegen sein. Der durch Adjunktion der Koordinaten von $\mathfrak{p}_0$ zu $\mathfrak{K}$ entstehende Körper werde mit $\mathfrak{K}(\mathfrak{p}_0)$ bezeichnet. Liegen die Koeffizienten a, b, c von L in $\mathfrak{K}$, so können die Koeffizienten von ψ in $\mathfrak{K}(\mathfrak{p}_0)$ gewählt werden. Die Ungleichung (85) gilt dann auch in den zu $\mathfrak{K}(\mathfrak{p}_0)$ in bezug auf $\mathfrak{K}$ konjugierten Körpern. Bedeutet r den Relativgrad von $\mathfrak{K}(\mathfrak{p}_0)$ und $\mathfrak{p}_0, \mathfrak{p}_0', \cdots, \mathfrak{p}_0^{(r-1)}$ die Konjugierten von $\mathfrak{p}_0$ in bezug auf $\mathfrak{K}$, ebenso $\mathfrak{q}_1, \mathfrak{q}_1', \cdots, \mathfrak{q}_1^{(r-1)}, \cdots$ die Konjugierten von $\mathfrak{q}_1, \cdots$, so ist das Produkt

$$(86) \qquad (\omega(\mathfrak{p}, \mathfrak{p}_0)\omega(\mathfrak{p}, \mathfrak{p}_0')\cdots)^{hm-p}(\omega(\mathfrak{p}, \mathfrak{q}_1)\omega(\mathfrak{p}, \mathfrak{q}_1')\cdots)(\omega(\mathfrak{p}, \mathfrak{q}_2)\omega(\mathfrak{p}, \mathfrak{q}_2')\cdots)\cdots$$

in $\mathfrak{K}$ gelegen. Endlich gehe man in (85) zu den zu $\mathfrak{K}$ in bezug auf den Körper der rationalen Zahlen konjugierten Körpern über; dann ist die Norm des Ausdrucks (86) kleiner als

$$c_1 N(|x|+|y|+|z|)^{mr},$$

wo c_1 von $\mathfrak{p}_0$ und m abhängt. Erst recht ist also

$$(87) \qquad N|\omega(\mathfrak{p}, \mathfrak{p}_0)\cdots\omega(\mathfrak{p}, \mathfrak{p}_0^{(r-1)})| < c_2 N(|x|+|y|+|z|)^{\frac{mr}{hm-p}},$$

mit analoger Bedeutung von c_2; dabei verlangt das Zeichen N, daß das Produkt über die zu $\mathfrak{K}$ konjugierten Körper erstreckt wird. In (87) kann m beliebig groß sein; es gilt demnach auch

$$(88) \qquad N|\omega(\mathfrak{p}, \mathfrak{p}_0)\cdots\omega(\mathfrak{p}, \mathfrak{p}_0^{(r-1)})| < c_3 N(|x|+|y|+|z|)^{\frac{r}{h}+\varepsilon}$$

für jedes $\varepsilon > 0$, wo c_3 von ε und $\mathfrak{p}_0$ abhängt.

Man betrachte zwei Formen χ_1, χ_2 gleichen Grades mit Koeffizienten aus $\mathfrak{K}$, deren gemeinschaftliche Nullstellen $\mathfrak{p}_1, \mathfrak{p}_2, \cdots$ seien. Eine beliebige Form der Schar $\lambda_1\chi_1+\lambda_2\chi_2$ hat dann außer $\mathfrak{p}_1, \cdots$ noch gewisse weitere Nullstellen $\mathfrak{r}, \cdots$. Es seien λ_1, λ_2 rationale Zahlen. Ersetzt man die Koeffizienten von χ_1 und χ_2 simultan durch ihre Konjugierten, so mögen die Nullstellen $\mathfrak{r}', \cdots; \mathfrak{r}'', \cdots; \cdots$ an die Stelle von $\mathfrak{r}, \cdots$ treten; man bezeichne die Gesamtheit der so entstehenden Nullstellen $\mathfrak{r}, \cdots; \mathfrak{r}', \cdots; \cdots$ mit $\mathfrak{S}$. Für jedes l kann man $l+1$ Formen der Schar angeben, etwa $\psi_0, \cdots, \psi_l$, so daß keine zwei der zugehörigen Nullstellensysteme $\mathfrak{S}_0, \cdots, \mathfrak{S}_l$ eine gemeinsame Nullstelle enthalten; man hat ja zur Konstruktion solcher Formen nur zu beachten, daß man λ_1, λ_2 sicherlich derart wählen kann, daß $\lambda_1\chi_1+\lambda_2\chi_2$ in gewissen endlich vielen Punkten nicht 0 ist, in denen χ_1, χ_2 nicht beide verschwinden. Sind nun $\mathfrak{q}_1, \cdots, \mathfrak{q}_l$ irgend l Punkte, so kommt $\mathfrak{q}_1$ in höchstens einem der Systeme $\mathfrak{S}_0, \cdots, \mathfrak{S}_l$ vor, etwa in $\mathfrak{S}_0$ und dann nicht in $\mathfrak{S}_1, \cdots, \mathfrak{S}_l$; ebenso kommt $\mathfrak{q}_2$ in höchstens einem der Systeme $\mathfrak{S}_1, \cdots, \mathfrak{S}_l$ vor, etwa in $\mathfrak{S}_1$ und dann nicht in $\mathfrak{S}_2, \cdots, \mathfrak{S}_l$; $\cdots$; also gibt es ein System, in welchem keiner der Punkte $\mathfrak{q}_1, \cdots, \mathfrak{q}_l$ auftritt.

Nun sei Φ eine Funktion des durch $\frac{x}{z}$ und $\frac{y}{z}$ erzeugten algebraischen Funktionenkörpers, mit Koeffizienten aus $\mathfrak{K}$. Ihre Nullstellen seien $\mathfrak{p}_1, \cdots, \mathfrak{p}_g$; ihre Pole seien $\mathfrak{q}_1, \cdots, \mathfrak{q}_g$. Es sei l der Grad des Körpers $\mathfrak{K}$ und $\mathfrak{p}$ wie oben in $\mathfrak{K}$ gelegen. Nach dem soeben Bewiesenen kann man aus $l+1$ gewissen Formen $\psi_0, \cdots, \psi_l$, die durch $\mathfrak{p}_1, \cdots, \mathfrak{p}_g$ bestimmt sind, eine Funktion ψ auswählen, welche in $\mathfrak{p}_1, \cdots, \mathfrak{p}_g$ verschwindet, deren sonstige Nullstellen $\mathfrak{r}, \cdots$ und Konjugierte dieser Nullstellen aber nicht in der Nähe von $\mathfrak{p}$ und den Konjugierten von $\mathfrak{p}$ gelegen sind; ferner sind die Koeffizienten dieser Form ψ Zahlen aus $\mathfrak{K}$. Dann ist $\psi : \Phi = \chi$ eine Form, die in den Punkten $\mathfrak{q}_1, \cdots, \mathfrak{q}_g, \mathfrak{r}, \cdots$ verschwindet, nirgendwo unendlich wird und denselben Grad γ besitzt wie ψ. Liegt nun $\mathfrak{p}$ nicht in der Umgebung der Nullstellen $\mathfrak{q}_1, \cdots, \mathfrak{q}_g, \mathfrak{r}, \cdots$ von χ, so gilt analog zu (85) die Abschätzung

$$|\chi| > c(|x|+|y|+|z|)^{\gamma},$$

also

$$c|\Phi|(|x|+|y|+|z|)^{\gamma} < |\psi|,$$

wo $c > 0$ noch von der Größe der für $\mathfrak{p}$ verbotenen Umgebungen abhängt. Diese Ungleichung gilt auch in den konjugierten Körpern, wenn für $\mathfrak{p}$ auch noch die Nähe der Nullstellen der zu χ konjugierten Funktionen ausgeschlossen wird. Also ist auch

$$N|\Phi|N(|x|+|y|+|z|)^{\gamma} < c_4 N|\omega(\mathfrak{p}, \mathfrak{p}_1)\cdots\omega(\mathfrak{p}, \mathfrak{p}_g)|N|\omega(\mathfrak{p}, \mathfrak{r})\cdots|.$$

Für den Faktor $N|\omega(\mathfrak{p}, \mathfrak{r})\cdots|$ benutze man die Abschätzung (88), wobei zu beachten ist, daß die Form ψ genau $h\gamma$ Nullstellen hat, die Anzahl der Nullstellen $\mathfrak{r}, \cdots$ also gleich $h\gamma - g$ ist; es wird

$$N|\omega(\mathfrak{p}, \mathfrak{r})\cdots| < c_5 N(|x|+|y|+|z|)^{\frac{h\gamma-g}{h}+\varepsilon}$$

(89) $$N|\Phi| < c_6 N|\omega(\mathfrak{p}, \mathfrak{p}_1)\cdots\omega(\mathfrak{p}, \mathfrak{p}_g)|N(|x|+|y|+|z|)^{-\frac{g}{h}+\varepsilon}.$$

Dies gilt unter der Voraussetzung, daß die Konjugierten von $\mathfrak{p}$ nicht in der Nähe der Konjugierten der Pole $\mathfrak{q}_1, \cdots, \mathfrak{q}_g$ von Φ liegen.

Die Ungleichung (89) bietet nun die Möglichkeit, aus der Annahme, eine Funktion des durch $\frac{x}{z}$ und $\frac{y}{z}$ erzeugten algebraischen Funktionenkörpers sei ganzzahlig für unendlich viele in $\mathfrak{K}$ gelegene Werte von $\frac{x}{z}$ und $\frac{y}{z}$, zu einer Aussage über Approximation einer gewissen algebraischen Zahl zu gelangen. Es sei nämlich F eine solche Funktion mit Koeffizienten aus $\mathfrak{K}$, ihre Ordnung sei g, ihre Pole seien $\mathfrak{p}_1, \cdots, \mathfrak{p}_g$, ihre Nullstellen seien $\mathfrak{q}_1, \cdots, \mathfrak{q}_g$. Es möge $\mathfrak{p}$ die Punkte durchlaufen, in denen F ganzzahlig ist. Da es zwei Formen gibt, die keine Nullstelle gemeinsam haben, von denen aber die eine in $\mathfrak{p}_1, \cdots, \mathfrak{p}_g$, die andere in $\mathfrak{q}_1, \cdots, \mathfrak{q}_g$ verschwindet, so haben $\omega(\mathfrak{p}, \mathfrak{p}_1)\cdots\omega(\mathfrak{p}, \mathfrak{p}_g)$ und $\omega(\mathfrak{p}, \mathfrak{q}_1)\cdots\omega(\mathfrak{p}, \mathfrak{q}_g)$ nur einen unitären Teiler gemeinsam. Andererseits ist der Quotient dieser beiden Zahlen bis auf einen unitären Faktor gleich der ganzen Zahl F. Folglich ist der Ausdruck $\omega(\mathfrak{p}, \mathfrak{p}_1)\cdots\omega(\mathfrak{p}, \mathfrak{p}_g)$ selbst unitär. Indem man nötigenfalls F durch eine der Funktionen $F, F+1, \cdots, F+l$ ersetzt, kann man erreichen, daß $\mathfrak{p}$ nicht in der Nähe einer Nullstelle von F gelegen ist und gleiches für alle l Konjugierten von $\mathfrak{p}$ und F gilt. Nun wende man (89) mit $\Phi = 1 : F$ an; es wird

(90) $$N|\Phi| < c_7 N(|x|+|y|+|z|)^{-\frac{g}{h}+\varepsilon}.$$

Für die unendlich vielen $\mathfrak{p}$, die F ganzzahlig machen, ist eine bestimmte der l Konjugierten von Φ unendlich oft absolut genommen am kleinsten; dies sei etwa Φ selber. Die linke Seite von (90) ist dann nicht kleiner als $|\Phi|^l$. Ferner ist leicht zu sehen, daß die rechte Seite gegen 0 konvergiert. Sind nämlich $\frac{x}{z}, \frac{x'}{z'}, \cdots$ die l Konjugierten von $\frac{x}{z}$, so ist

$$(zt-x)(z't-x')\cdots = 0$$

eine Gleichung l-ten Grades mit ganzen rationalen Koeffizienten für $t = \frac{x}{z}$; und offenbar sind diese Koeffizienten absolut genommen kleiner als $c_8 N(|x|+|z|)$. Daher ist $c_8 N(|x|+|y|+|z|)$ eine obere Schranke für die absoluten Beträge der ganzen rationalen Koeffizienten in den Gleichungen, denen $\frac{x}{z}$ und $\frac{y}{z}$ genügen. Wegen der Existenz unendlich vieler $\mathfrak{p}$ wird also $N(|x|+|y|+|z|)$ unendlich. Aus (90) ergibt sich nun die Konvergenz einer Teilfolge der $\mathfrak{p}$ gegen eine feste Nullstelle von Φ, etwa $\mathfrak{p}_1$. Dies sei eine Nullstelle der Ordnung r. Es bedeute φ irgendeine in $\mathfrak{p}_1$ verschwindende rationale

Funktion von $\frac{x}{z}$ und $\frac{y}{z}$. Dann ist $\phi^r : \Phi$ für $\mathfrak{p} \to \mathfrak{p}_r$ beschränkt. Damit ist das für alles weitere wichtige Resultat gewonnen:

Es sei $f\left(\frac{x}{z}, \frac{y}{z}\right) = 0$ die Gleichung einer algebraischen Kurve h-ter Ordnung mit Koeffizienten aus einem algebraischen Zahlkörper $\mathfrak{K}$ vom Grade l. Es gebe eine Funktion g-ter Ordnung aus dem durch $\frac{x}{z}$ und $\frac{y}{z}$ erzeugten algebraischen Funktionenkörper mit Koeffizienten aus $\mathfrak{K}$, deren Wert für unendlich viele in $\mathfrak{K}$ gelegene Kurvenpunkte $\frac{x}{z}, \frac{y}{z}$ ganzzahlig ist. Dann konvergiert eine Teilfolge dieser Kurvenpunkte gegen einen Pol der Funktion. Es sei r die Ordnung dieses Poles. Dann gilt für jede in dem Pole verschwindende Funktion ϕ des Funktionenkörpers die Ungleichung

$$(91) \qquad |\phi| < c_9 N(|x|+|y|+|z|)^{-\frac{g}{hlr}+\varepsilon},$$

falls $\frac{x}{z}, \frac{y}{z}$ die Teilfolge durchläuft, die Zahlen x, y, z teilerfremd gewählt werden und ε eine beliebig kleine positive Zahl bedeutet; die Zahl c_9 hängt noch von ε ab, jedoch nicht von x, y, z.

Dies ist ein Approximationssatz für eine Nullstelle von ϕ. Für seine Anwendung ist es wichtig, die Zahl g hinreichend groß wählen zu können. Dies wird ermöglicht durch die Teilungstheorie der Abelschen Funktionen in Verbindung mit dem zu Anfang genannten Weilschen Satz. Zunächst sollen noch die Kurven vom Geschlecht 1 behandelt werden, bei denen gewisse Schwierigkeiten des allgemeinen Falles nicht auftreten.

§ 3. Gleichungen vom Geschlecht 1.

Ist die Gleichung $f(x, y) = 0$ vom Geschlecht 1, so geht sie durch eine birationale Transformation

$$(92) \qquad x = \phi(u, t), \qquad y = \psi(u, t)$$

über in die Gleichung

$$t^2 = 4u^3 - g_2 u - g_3.$$

Die Koeffizienten der rationalen Funktionen ϕ und ψ sowie die Größen g_2, g_3 sind algebraische Zahlen; ohne Beschränkung der Allgemeinheit kann angenommen werden, daß sie in dem Körper $\mathfrak{K}$ enthalten sind, da man diesen sonst nur zu erweitern brauchte. Bedeutet $\wp(s)$ die mit den Invarianten g_2, g_3 gebildete Weierstrasssche $\wp$-Funktion, so wird die Kurve $f = 0$ durch den Ansatz

$$(93) \qquad t = \wp(s), \qquad u = \wp'(s)$$

uniformisiert. Der Körper der zu den Invarianten g_2, g_3 gehörigen elliptischen Funktionen stimmt überein mit dem durch x und y erzeugten algebraischen Funktionenkörper. Es sei $w(s)$ irgendeine nicht konstante Funktion dieses Körpers und r ihre Ordnung, also die Anzahl der Male, die sie im Periodenparallelogramm jeden beliebigen Wert a annimmt. Es seien $v_1, \cdots, v_r$ sämtliche im Periodenparallelogramm gelegenen Lösungen von $w(s) = a$, und zwar jede mit ihrer Vielfachheit hingeschrieben. Bedeutet n eine natürliche Zahl und c eine beliebige Zahl, so liegen die a-Stellen der elliptischen Funktion $w(ns+c)$ genau in den Punkten $s = \frac{1}{n}(v_k - c + \omega)$, wo $k = 1, \cdots, r$ und ω eine beliebige Periode ist. Innerhalb des Periodenparallelogramms entstehen daher aus jeder l-fachen a-Stelle von $w(s)$ genau n^2 voneinander verschiedene l-fache a-Stellen von $w(ns+c)$; dabei führen verschiedene a-Stellen von $w(s)$ auch zu verschiedenen a-Stellen von $w(ns+c)$. Folglich ist die Funktion $w(ns+c)$ von der Ordnung $n^2 r = g$ und nimmt keinen Wert mehr als r-fach an.

Man betrachte nun sämtliche in $\mathfrak{K}$ gelegenen Lösungen von $f(x, y) = 0$ in ganzen oder gebrochenen x, y. Nach dem Satze von Mordell und Weil bilden die zu diesen

x, y gehörigen Werte des Integrals erster Gattung s einen Modul $\mathfrak{M}$ von endlicher Basis. Es seien $s_1, \cdots, s_q$ die Basiselemente. Dann erhält man alle in $\mathfrak{K}$ gelegenen Lösungen von $f = 0$ aus (92) und (93), wenn darin

$$s = n_1 s_1 + \cdots + n_q s_q \tag{94}$$

gesetzt wird und $n_1, \cdots, n_q$ alle ganzen rationalen Zahlen durchlaufen. Es sei n eine natürliche Zahl. Nach (94) hat dann jedes Element von $\mathfrak{M}$ die Gestalt

$$s = n\sigma + c, \tag{95}$$

wo σ ein Element von $\mathfrak{M}$ und c eins von endlich vielen Elementen von $\mathfrak{M}$ ist; es genügt nämlich, c auf die Werte $n_1 s_1 + \cdots + n_q s_q$ mit $n_k = 0, \cdots, n-1$ $(k = 1, \cdots, q)$ zu beschränken.

Jetzt werde angenommen, $f = 0$ habe sogar unendlich viele Lösungen aus $\mathfrak{K}$ mit ganzzahligem x. Aus diesen greife man unendlich viele heraus, für welche die Zahl c in (95) einen und denselben festen Wert hat. Nun identifiziere man x mit der elliptischen Funktion $w(s) = w(n\sigma + c)$. Gehört zu σ die Lösung ξ, η von $f = 0$, so ist nach dem Additionstheorem x eine rationale Funktion von ξ, η mit Koeffizienten aus $\mathfrak{K}$, deren Ordnung den Wert $n^2 r = g$ besitzt und deren g Pole höchstens r-fach sind. Die Pole bestimmen sich durch Lösung einer algebraischen Gleichung g-ten Grades mit Koeffizienten aus $\mathfrak{K}$. Nach dem Ergebnis des vorigen Paragraphen wird einer der Pole durch die in $\mathfrak{K}$ gelegenen unendlich vielen Zahlenpaare ξ, η approximiert. Man ersetze ξ, η durch $\frac{\xi}{\zeta}, \frac{\eta}{\zeta}$ mit teilerfremden ξ, η, ζ.

Konvergiert $\frac{\xi}{\zeta}$ gegen eine endliche Zahl ρ, so ist ρ höchstens vom Grade g in bezug auf $\mathfrak{K}$ und es gilt nach (91)

$$\left|\frac{\xi}{\zeta} - \rho\right| < c_9 N(|\xi| + |\zeta|)^{-\varkappa g + 1}; \tag{96}$$

dabei ist $\varkappa = 1 : hlr$, wo h die Ordnung der Kurve $f = 0$, l den Grad des Körpers $\mathfrak{K}$ und r den Grad von f in y bedeuten; ferner ist $g = n^2 r$ mit beliebigem natürlichen n. Wie bereits früher bemerkt wurde, wächst $N(|\xi| + |\zeta|)$ mindestens ebenso stark an wie der größte der absoluten Beträge der ganzen rationalen Koeffizienten in der Gleichung l-ten Grades für $\frac{\xi}{\zeta}$, der mit $H\left(\frac{\xi}{\zeta}\right)$ bezeichnet werde. Nach der erwähnten Verallgemeinerung des Thueschen Satzes ist nun andererseits

$$\left|\frac{\xi}{\zeta} - \rho\right| > c_{10}\left\{H\left(\frac{\xi}{\zeta}\right)\right\}^{-\lambda\sqrt{g}}, \tag{97}$$

wo λ nur vom Grade des Körpers $\mathfrak{K}$ abhängt, dem die approximierende Zahl $\xi : \zeta$ angehört. Für hinreichend großes n, nämlich für $n > \lambda h l \sqrt{r}$, können aber (96) und (97) sicherlich nur durch endlich viele $\xi : \zeta$ erfüllt werden.

Konvergiert aber $\xi : \zeta$ gegen ∞, so hat man in der soeben ausgeführten Überlegung nur $\zeta : \xi$ und 0 an Stelle von $\xi : \zeta$ und ρ zu nehmen und kommt zum gleichen Resultat.

Die Übertragung des Beweises auf den allgemeinen Fall $p \geq 1$ erfordert zunächst einige Hilfssätze über die Teilung der Abelschen Funktionen.

§ 4. Hilfsmittel aus der Theorie der Abelschen Funktionen.

Es sei $\mathfrak{R}$ eine Riemannsche Fläche vom Geschlecht $p \geq 1$. Sie werde kanonisch zerschnitten durch ein System von p Rückkehrschnittpaaren $\mathfrak{A}_l, \mathfrak{B}_l$ $(l = 1, \cdots, p)$. Es sei $\mathfrak{p}$ ein variabler Punkt der Fläche. Wie in der Einleitung bedeute $w_1(\mathfrak{p}), \cdots, w_p(\mathfrak{p})$ ein System von Normalintegralen erster Gattung. Die zu $\mathfrak{A}_l$ gehörige Periode von w_k ist dann e_{kl}, d. h. $= 0$ oder $= 1$, je nachdem $k \neq l$ oder $k = l$ ist; die zu $\mathfrak{B}_l$ gehörige

Periode von w_k sei τ_{kl}. Auf der unzerschnittenen Fläche sind $w_1, \cdots, w_p$ nur bis auf die Perioden $\Omega_1, \cdots, \Omega_p$ bestimmt, wo

$$(98) \qquad \Omega_k = \sum_{l=1}^{p} (g_l e_{kl} + h_l \tau_{kl}) \qquad (k = 1, \cdots, p)$$

gesetzt ist und $g_1, \cdots, g_p, h_1, \cdots, h_p$ ganze rationale Zahlen bedeuten. Zwei Systeme von p Zahlen $a_1, \cdots, a_p$ und $b_1, \cdots, b_p$, für welche die Differenzen $a_1 - b_1, \cdots, a_p - b_p$ gleich einem System simultaner Perioden $\Omega_1, \cdots, \Omega_p$ sind, mögen als kongruent bezeichnet werden.

Es sei

$$\vartheta(s) = \vartheta(s_1, \cdots, s_p) = \sum_{n_1=-\infty}^{+\infty} \cdots \sum_{n_p=-\infty}^{+\infty} e^{\pi i \sum_{k,l} \tau_{kl} n_k n_l + 2\pi i \sum_k n_k s_k}$$

die RIEMANNsche Thetafunktion mit den unabhängigen Variabeln $s_1, \cdots, s_p$. Es gibt $p + 1$ feste Punkte $\mathfrak{a}, \mathfrak{a}_1, \cdots, \mathfrak{a}_p$ von der Art, daß das Umkehrproblem

$$(99) \qquad \sum_{l=1}^{p} \{ w_k(\mathfrak{p}_l) - w_k(\mathfrak{a}_l) \} \equiv s_k \qquad (k = 1, \cdots, p)$$

dann und nur dann durch genau eine Punktgruppe $\mathfrak{p}_1, \cdots, \mathfrak{p}_p$ gelöst werden kann, wenn die Thetafunktion mit den Argumenten $s_k - w_k(\mathfrak{p}) + w_k(\mathfrak{a})$, also $\vartheta(s - w(\mathfrak{p}) + w(\mathfrak{a}))$, nicht identisch in $\mathfrak{p}$ verschwindet. Die Punktgruppe $\mathfrak{p}_1, \cdots, \mathfrak{p}_p$ werde kurz mit $\mathfrak{P}$ bezeichnet. Es sei $\pi_1, \cdots, \pi_p$ irgendeine andere Punktgruppe Π; man setze analog zu (99)

$$(100) \qquad \sum_{l=1}^{p} \{ w_k(\pi_l) - w_k(\mathfrak{a}_l) \} \equiv \sigma_k. \qquad (k = 1, \cdots, p)$$

Es seien ferner n eine natürliche Zahl und $c_1, \cdots, c_p$ irgendwelche Konstanten; dann wird durch die Forderung

$$(101) \qquad s_k \equiv n\sigma_k + c_k \qquad (k = 1, \cdots, p)$$

eine Beziehung zwischen $\mathfrak{P}$ und Π festgelegt. Aus (101) folgt

$$(102) \qquad \sigma_k \equiv \frac{1}{n}(s_k - c_k) + \frac{1}{n}\Omega_k,$$

wo in dem Ausdruck (98), welcher Ω_k definiert, für $g_1, \cdots, g_p, h_1, \cdots, h_p$ unabhängig voneinander die Zahlen $0, \cdots, n-1$ zu setzen sind; diese führen nämlich zu allen n^{2p} inkongruenten Systemen $\frac{1}{n}\Omega_1, \cdots, \frac{1}{n}\Omega_p$. Durch (102) erhält man also aus dem einen System $s_1, \cdots, s_p$ insgesamt n^{2p} inkongruente Systeme $\sigma_1, \cdots, \sigma_p$.

Man betrachte nun speziell die Punktgruppe $\mathfrak{P}_0$, die aus dem p-mal gezählten Punkt $\mathfrak{p}$ besteht. Das zugehörige s_k ist dann

$$s_k \equiv p\, w_k(\mathfrak{p}) - \sum_{l=1}^{p} w_k(\mathfrak{a}_l); \qquad (k = 1, \cdots, p)$$

und nach (102) wird

$$(103) \qquad \sigma_k \equiv \frac{p}{n} w_k(\mathfrak{p}) + b_k$$

mit

$$b_k = \frac{1}{n}\left(-\sum_{l=1}^{p} w_k(\mathfrak{a}_l) - c_k + \Omega_k \right).$$

Es soll jetzt gezeigt werden, daß für alle hinreichend großen n den Kongruenzen (100) und (103) durch genau eine Punktgruppe Π genügt wird, falls von endlich vielen Ausnahmepunkten $\mathfrak{p}$ abgesehen wird. Zu diesem Zwecke hat man zu zeigen, daß nur für endlich viele $\mathfrak{p}$ die Funktion

$$(104) \qquad \vartheta\left(\frac{p}{n} w(\mathfrak{p}) + b - w(\mathfrak{q}) + w(\mathfrak{a}) \right)$$

identisch in $\mathfrak{q}$ verschwindet. Würde dies für unendlich viele $\mathfrak{p}$ gelten, so gälte es auch identisch in $\mathfrak{p}$, denn die Funktion (104) ist eine auf $\mathfrak{R}$ ausnahmslos reguläre Funktion des Punktes $\mathfrak{p}$. Bei Umläufen von $\mathfrak{p}$ auf $\mathfrak{R}$ vermehrt sich $w(\mathfrak{p})$ um eine Periode Ω; das Argument in (104) vermehrt sich dabei um $\frac{p}{n}\Omega$. Läßt man $\Omega_1, \cdots, \Omega_p$ alle Periodensysteme durchlaufen und n über alle Grenzen wachsen, so konvergieren die Werte $\frac{p}{n}\Omega_1, \cdots, \frac{p}{n}\Omega_p$ gegen jedes System von komplexen Zahlen. Nun verschwindet aber doch $\vartheta(s)$ nicht identisch als Funktion der unabhängigen Variabeln $s_1, \cdots, s_p$; folglich kann (104) für hinreichend großes n nicht identisch in $\mathfrak{p}$ verschwinden. Dies gilt offenbar unabhängig von der Wahl der Konstanten $b_1, \cdots, b_p$. Fortan sei n hinreichend groß.

Wird von endlich vielen $\mathfrak{p}$ abgesehen, so sind für jede Punktgruppe $\mathfrak{P}_0$ vermöge (101) genau n^{2p} verschiedene Punktgruppen Π bestimmt; es sei $\pi_1, \cdots, \pi_p$ eine von ihnen. Es bedeute $\chi(\mathfrak{p})$ eine nicht konstante rationale Funktion auf der Fläche $\mathfrak{R}$; ihre Ordnung sei ρ, ihre Nullstellen seien $\nu^{(1)}, \cdots, \nu^{(\rho)}$, ihre Pole seien $\pi^{(1)}, \cdots, \pi^{(\rho)}$. Dann ist

$$(105) \qquad \chi(\pi_1) \cdots \chi(\pi_p) = c \prod_{m=1}^{\rho} \frac{\vartheta(\sigma - w(\nu^{(m)}) + w(\mathfrak{a}))}{\vartheta(\sigma - w(\pi^{(m)}) + w(\mathfrak{a}))}$$

mit konstantem c. Läßt man $\mathfrak{p}$ auf $\mathfrak{R}$ eine geschlossene Kurve n-mal durchlaufen, so vermehrt sich nach (103) der Wert σ_k um eine volle Periode. Die Abelschen Funktionen von $\sigma_1, \cdots, \sigma_p$ sind daher als Funktionen der Variabeln $\mathfrak{p}$ eindeutig rational auf derjenigen Überlagerungsfläche $\mathfrak{U}$ der Riemannschen Fläche $\mathfrak{R}$, auf welcher alle n-mal durchlaufenen, auf $\mathfrak{R}$ geschlossenen Kurven wieder geschlossen sind. Um $\mathfrak{U}$ aus $\mathfrak{R}$ zu erhalten, hat man die n^{2p} Decktransformationen auf $\mathfrak{R}$ auszuführen, welche aus g_l-maligem Durchlaufen der Rückkehrschnitte $\mathfrak{A}_l$ und h_l-maligem Durchlaufen der Rückkehrschnitte $\mathfrak{B}_l$ $(l = 1, \cdots, p;$ $g_l = 0, \cdots, n-1; h_l = 0, \cdots, n-1)$ entstehen. Nun soll die Ordnung der speziellen Funktion $\chi(\pi_1) \cdots \chi(\pi_p) = \Phi(\mathfrak{p})$ auf $\mathfrak{U}$ bestimmt werden. Zu diesem Zwecke hat man die Anzahl der auf $\mathfrak{U}$ gelegenen Nullstellen der Funktion (104) zu ermitteln. Man bilde gleich etwas allgemeiner $\vartheta\left(\frac{q}{n}w(\mathfrak{p}) + a\right)$, wo q eine natürliche Zahl, $a_1, \cdots, a_p$ irgendwelche Konstanten bedeuten. Die Anzahl der Nullstellen dieser Funktion ϑ ist gleich der Änderung von $\frac{1}{2\pi i}\log\vartheta$ bei positivem Umlaufen des Randes der kanonisch zerschnittenen Fläche $\mathfrak{U}$. Da sich $\mathfrak{U}$ aus n^{2p} Exemplaren von $\mathfrak{R}$ zusammensetzt, die durch die obengenannten Decktransformationen auseinander hervorgehen, so hat man nur die Summe der Änderungen von $\frac{1}{2\pi i}\log\vartheta$ beim Umlaufen des Randes der n^{2p} Exemplare der kanonisch zerschnittenen Fläche $\mathfrak{R}$ zu berechnen. Beachtet man die Gleichungen

$$\vartheta(s + e_k) = \vartheta(s), \qquad \vartheta(s + \tau_k) = e^{-\pi i \tau_{kk} - 2\pi i s_k}\vartheta(s),$$

also

$$\vartheta\left(\frac{q}{n}s + qe_k\right) = \vartheta\left(\frac{q}{n}s\right), \qquad \vartheta\left(\frac{q}{n}s + q\tau_k\right) = e^{-\pi i q^2 \tau_{kk} - 2\pi i \frac{q^2}{n} s_k}\vartheta\left(\frac{q}{n}s\right),$$

so folgt in üblicher Weise, daß die Schnitte $\mathfrak{B}_l$ keinen Beitrag liefern, während n^{2p-2}-mal von den Schnitten $\mathfrak{A}_l$ der Beitrag $\frac{q^2}{n}\cdot n$ geliefert wird, und zwar für $l = 1, \cdots, p$. Die Anzahl der Nullstellen ist somit genau $p\cdot q^2 n^{2p-2}$. Da die rechte Seite des Ausdrucks (105) für $\Phi(\mathfrak{p})$ einen Faktor ϑ genau ρ-mal im Nenner enthält, *so ist die Ordnung von* $\Phi(\mathfrak{p})$ *auf* $\mathfrak{U}$ *höchstens gleich* $\rho p^3 n^{2p-2}$. Für den Zweck der Abhandlung ist wesentlich, daß der Exponent von n um nicht weniger als zwei kleiner ist als $2p$.

Man setze speziell

$$(106) \qquad \chi(\mathfrak{p}) = t - \frac{\alpha_1 x + \beta_1 y + \gamma_1}{\alpha_2 x + \beta_2 y + \gamma_2}$$

mit irgendwelchen Konstanten $t, \alpha_1, \beta_1, \gamma_1, \alpha_2, \beta_2, \gamma_2$, von denen $\alpha_1, \beta_1, \gamma_1$ nicht zu $\alpha_2, \beta_2, \gamma_2$ proportional sind; dabei sollen x und y zwei Funktionen bedeuten, die den zu $\mathfrak{R}$ gehörigen algebraischen Funktionenkörper erzeugen, also Koordinaten des Punktes $\mathfrak{p}$. Bedeutet h den Grad der zwischen x und y bestehenden algebraischen Gleichung, so ist die Ordnung ρ von $\chi(\mathfrak{p})$ auf $\mathfrak{R}$ gleich h. Die Ordnung von $\chi(\pi_1)\cdots\chi(\pi_p) = \Phi(\mathfrak{p})$ auf $\mathfrak{U}$ ist also höchstens gleich hp^3n^{2p-2}. In (106) wähle man p verschiedene Werte $t_1, \cdots, t_p$ für t; dadurch gehe $\Phi(\mathfrak{p})$ über in

$$\Phi_k(\mathfrak{p}) = t_k^p + C_1 t_k^{p-1} + \cdots + C_p. \qquad (k = 1, \cdots, p) \tag{107}$$

Eine zweite für das Folgende wichtige Tatsache ist nun:

Zwischen $\Phi_1(\mathfrak{p}), \cdots, \Phi_p(\mathfrak{p})$ besteht identisch in $\mathfrak{p}$ keine algebraische Gleichung, deren Grad kleiner ist als $\dfrac{n^2}{hp^{2p+1}}$.

Sind $\nu_k^{(1)}, \cdots, \nu_k^{(h)}$ die t_k-Stellen von $\dfrac{\alpha_1 x + \beta_1 y + \gamma_1}{\alpha_2 x + \beta_2 y + \gamma_2}$, so ist nach (105)

$$\Phi_k(\mathfrak{p}) = c\prod_{m=1}^{h} \frac{\vartheta(\sigma - w(\nu_k^{(m)}) + w(\mathfrak{a}))}{\vartheta(\sigma - w(\pi^{(m)}) + w(\mathfrak{a}))}. \qquad (k = 1, \cdots, p) \tag{108}$$

Es bestehe nun zwischen den rechten Seiten von (108) eine algebraische Gleichung vom Grade δ identisch in $\mathfrak{p}$; dabei sind $\sigma_1, \cdots, \sigma_p$ die durch (103) bestimmten Funktionen der einen Variabeln $\mathfrak{p}$. Die Gleichung werde kurz mit $G(\sigma) = 0$ bezeichnet. Sie ist nicht erfüllt, wenn $\sigma_1, \cdots, \sigma_p$ unabhängige Variable bedeuten; denn dann wären auch $\pi_1, \cdots, \pi_p$ unabhängige Variable, also auch $C_1, \cdots, C_p$ in (107), also auch $\Phi_1, \cdots, \Phi_p$ selber. Setzt man

$$\sigma_k \equiv \frac{1}{n}(w_k(\mathfrak{p}_1) + \cdots + w_k(\mathfrak{p}_p)) + b_k, \qquad (k = 1, \cdots, p)$$

so ist $G(\sigma)$ identisch gleich 0, wenn die Punkte $\mathfrak{p}_1, \cdots, \mathfrak{p}_p$ alle zusammenfallen, aber nicht identisch gleich 0, wenn sie unabhängig voneinander sind. Nun sei q die kleinste natürliche Zahl von der Art, daß $G(\sigma)$ identisch verschwindet, wenn $\mathfrak{p}_1, \cdots, \mathfrak{p}_q$ unter der Bedingung $\mathfrak{p}_1 = \cdots = \mathfrak{p}_q$ und $\mathfrak{p}_{q+1}, \cdots, \mathfrak{p}_p$ beliebig veränderlich sind. Es ist $1 < q \leq p$. Man wähle dann

$$\sigma_k \equiv \frac{1}{n}((q-1)w_k(\mathfrak{p}) + w_k(\mathfrak{p}_q) + \cdots + w_k(\mathfrak{p}_p)) + b_k. \tag{109}$$

Für dieses Argument verschwindet $G(\sigma)$ nicht identisch in $\mathfrak{p}, \mathfrak{p}_q, \cdots, \mathfrak{p}_p$. Man betrachte $G(\sigma)$ als Funktion von $\mathfrak{p}_q$; sie ist auf $\mathfrak{U}$ eindeutig rational. Der Hauptnenner der in $G(\sigma)$ eingehenden Thetaquotienten ist

$$\left\{\prod_{m=1}^{h} \vartheta(\sigma - w(\pi^{(m)}) + w(\mathfrak{a}))\right\}^{\delta}.$$

Nach Früherem verschwindet dieser Ausdruck auf $\mathfrak{U}$ an genau δhpn^{2p-2} Stellen $\mathfrak{p}_q$. Andererseits ist $\vartheta(\sigma)$ identisch gleich 0, wenn in (109) die Variable $\mathfrak{p}_q$ gleich $\mathfrak{p}$ gesetzt wird. Bei Umläufen auf $\mathfrak{R}$ geht $(q-1)w_k(\mathfrak{p}) + w_k(\mathfrak{p})$ in $(q-1)w_k(\mathfrak{p}) + w_k(\mathfrak{p}) + q\Omega_k$ über. Daher verschwindet $G(\sigma)$ als Funktion von $\mathfrak{p}_q$ sicherlich in $\mathfrak{p}$ und den Punkten, die durch eine q-mal wiederholte Decktransformation aus $\mathfrak{p}$ hervorgehen. Ist d der größte gemeinsame Teiler von q und n, so hat also $G(\sigma)$ als Funktion von $\mathfrak{p}_q$ auf $\mathfrak{U}$ mindestens $(n:d)^{2p}$ Nullstellen. Folglich ist

$$\left(\frac{n}{d}\right)^{2p} \leq \delta hpn^{2p-2}$$

$$\delta \geq \frac{n^2}{d^{2p}hp}. \tag{110}$$

Wegen $d \leq q \leq p$ folgt die Behauptung. Für das Weitere ist wiederum wesentlich, daß der Exponent von n in (110) nicht kleiner als 2 ist.

§ 5. Gleichungen von beliebigem positiven Geschlecht.

Es sei $f(x, y) = 0$ vom Geschlecht $p \geq 1$. Man betrachte alle Systeme von p Punkten $x_1, y_1;\ x_2, y_2;\ \cdots;\ x_p, y_p$, deren sämtliche rationalen symmetrischen Verbindungen mit Koeffizienten aus $\mathfrak{K}$ wieder in $\mathfrak{K}$ liegen. Die diesen Punktgruppen nach (99) zugeordneten Systeme $s_1, \cdots, s_p$ bilden auf Grund des Weilschen Satzes einen endlichen Modul $\mathfrak{M}$. Es seien die Systeme $s_1^{(1)}, \cdots, s_p^{(1)}; \cdots; s_1^{(q)}, \cdots, s_p^{(q)}$ eine Basis des Moduls, dann ist analog zu (95)

$$s_k \equiv n\sigma_k + c_k, \qquad (k = 1, \cdots, p) \tag{111}$$

wo $\sigma_1, \cdots, \sigma_p$ ein Element des Moduls und $c_1, \cdots, c_p$ eines von endlich vielen Elementen von $\mathfrak{M}$ ist; es genügt, c_k auf die Werte $n_1 s_k^{(1)} + \cdots + n_q s_k^{(q)}$ zu beschränken, wo $n_1, \cdots, n_q$ Zahlen der Reihe $0, \cdots, n-1$ sind.

Nun möge $f(x, y) = 0$ unendlich viele Lösungen in Zahlen x, y aus $\mathfrak{K}$ besitzen, von denen x ganz ist. Der Punkt x, y werde mit $\mathfrak{p}$ bezeichnet. Die zur Punktgruppe $\mathfrak{p}, \cdots, \mathfrak{p}$ gehörigen Werte von $s_1, \cdots, s_p$ liegen in $\mathfrak{M}$. Aus der Menge der $\mathfrak{p}$ werde eine unendliche Teilmenge herausgegriffen, für welche in (111) ein festes System $c_1, \cdots, c_p$ auftritt. Auf diese $\mathfrak{p}$ bezieht sich alles Weitere. Schließt man noch endlich viele $\mathfrak{p}$ aus, so sind die n^{2p} Punktgruppen $\pi_1, \cdots, \pi_p$ eindeutig bestimmt. Es seien ξ_l, η_l die Koordinanten des Punktes π_l $(l = 1, \cdots, p)$. Nach (111) gibt es unter den n^{2p} Punktgruppen $\pi_1, \cdots, \pi_p$ eine, für welche alle rationalen symmetrischen Funktionen der p Paare ξ_l, η_l mit Koeffizienten aus $\mathfrak{K}$ einen wieder in $\mathfrak{K}$ gelegenen Wert besitzen. Wählt man für die Konstanten $\alpha_1, \cdots, \gamma_2$, $t_1, \cdots, t_p$ des vorigen Paragraphen Werte aus $\mathfrak{K}$, so liegen insbesondere die Größen $\Phi_1(\mathfrak{p}), \cdots, \Phi_p(\mathfrak{p})$ sämtlich in $\mathfrak{K}$.

Der Körper aller rationalen symmetrischen Funktionen der p Paare ξ_l, η_l ist enthalten in dem zu $\mathfrak{U}$ gehörigen algebraischen Funktionenkörper; seine Riemannsche Fläche ist also entweder $\mathfrak{U}$ selber oder eine Fläche $\mathfrak{U}'$, zu welcher $\mathfrak{U}$ Überlagerungsfläche ist. Es bestehe $\mathfrak{U}'$ aus ν Exemplaren von $\mathfrak{R}$; dann ist ν ein Teiler von n^{2p}. Es soll gezeigt werden, daß bei geeigneter Wahl der Konstanten $\alpha_1, \cdots, \gamma_2, t_1, \cdots, t_p$ die beiden Funktionen $\Phi_1(\mathfrak{p})$ und $\Phi_k(\mathfrak{p})$ für $k = 2, \cdots, p$ den zu $\mathfrak{U}'$ gehörigen Körper erzeugen. Zu jedem Punkte $\mathfrak{p}$ von $\mathfrak{U}'$ gehört genau eine Punktgruppe $\pi_1, \cdots, \pi_p$; zu verschiedenen $\mathfrak{p}$ auch verschiedene Systeme $\pi_1, \cdots, \pi_p$. Es ist

$$\Phi_k(\mathfrak{p}) = \prod_{l=1}^{p}\left(t_k - \frac{\alpha_1\xi_l + \beta_1\eta_l + \gamma_1}{\alpha_2\xi_l + \beta_2\eta_l + \gamma_2}\right) = t_k^p + C_1 t_k^{p-1} + \cdots + C_p. \qquad (k = 1, \cdots, p) \tag{112}$$

Es bedeute a einen Wert, den Φ_1 nicht mehrfach annimmt. Ist λ die Ordnung von Φ_1 auf $\mathfrak{U}'$, so genügen λ verschiedene Punkte $\mathfrak{p}$ der Gleichung $\Phi_1(\mathfrak{p}) = a$. Für diese λ Punkte betrachte man das System der Koeffizienten $C_1, \cdots, C_p$ auf der rechten Seite von (112); es kann vorkommen, daß in zwei verschiedenen der λ Punkte $\mathfrak{p}$ die Systeme $C_1, \cdots, C_p$ übereinstimmen. Dann müßte in diesen beiden $\mathfrak{p}$ auch das System der p Ausdrücke $(\alpha_1\xi_l + \beta_1\eta_l + \gamma_1):(\alpha_2\xi_l + \beta_2\eta_l + \gamma_2)$ $(l = 1, \cdots, p)$ übereinstimmen, abgesehen von der Reihenfolge. Dies kann aber nicht identisch in $\alpha_1, \cdots, \gamma_2$ gelten, denn sonst würden die beiden zugehörigen Punktsysteme ξ_l, η_l $(l = 1, \cdots, p)$ ebenfalls übereinstimmen. Man kann also $\alpha_1, \cdots, \gamma_2$ so wählen, daß für die λ Lösungen von $\Phi_1 = a$ die λ Systeme $C_1, \cdots, C_p$ verschieden sind; und nun wähle man t_k für $k = 2, \cdots, p$ derart, daß die Werte von $t_k^p + C_1 t_k^{p-1} + \cdots + C_p$ für diese λ Systeme $C_1, \cdots, C_p$ sämtlich verschieden ausfallen. Dann erzeugen Φ_1 und Φ_k den zu $\mathfrak{U}'$ gehörigen Körper. Nach § 4 ist die Ordnung von Φ_k und allgemeiner jeder mit konstanten Koeffizienten gebildeten linearen Kombination von $\Phi_1, \cdots, \Phi_p$ auf $\mathfrak{U}$ höchstens gleich hp^3n^{2p-2}; auf $\mathfrak{U}'$ ist diese Ordnung also höchstens gleich $(hp^3n^{2p-2}):\left(\frac{n^{2p}}{\nu}\right) = \frac{hp^3\nu}{n^2}$. Der Grad der zwischen Φ_1 und Φ_k bestehenden algebraischen Gleichung ist also höchstens gleich $hp^3\nu : n^2$. Die Koeffizienten dieser Gleichung liegen in $\mathfrak{K}$.

Es sei ζ_0 der Hauptnenner der p Zahlen $\Phi_1, \cdots, \Phi_p$, also $\Phi_1 = \zeta_1 : \zeta_0, \cdots, \Phi_p = \zeta_p : \zeta_0$ mit teilerfremden $\zeta_0, \zeta_1, \cdots, \zeta_p$. Ferner sei μ_k der Hauptnenner von Φ_1 und Φ_k für $k = 2, \cdots, p$. Man bilde den Ausdruck

$$(113)\qquad A = \prod_{k=2}^{p} N(|\mu_k \Phi_1| + |\mu_k \Phi_k| + |\mu_k|),$$

wo die Norm N über die l zu $\mathfrak{K}$ konjugierten Körper erstreckt wird. Offenbar ist das Produkt $\mu_2 \cdots \mu_p$ ein Vielfaches von ζ_0 und folglich

$$(114)\qquad \begin{aligned} A &\geq N(|\zeta_0| \prod_{k=2}^{p} (|\Phi_1| + |\Phi_k| + 1)) \\ &\geq N(|\zeta_0 \Phi_1| + |\zeta_0 \Phi_2| + \cdots + |\zeta_0 \Phi_p| + |\zeta_0|) = N(|\zeta_0| + \cdots + |\zeta_p|). \end{aligned}$$

Das Ergebnis von § 2 werde nun angewendet auf die Funktion x, die auf $\mathfrak{R}$, also auch auf $\mathfrak{U}'$ rational ist und daher dem durch Φ_1 und Φ_k erzeugten Funktionenkörper angehört. Nach Voraussetzung ist x ganzzahlig. Die Ordnung von x auf $\mathfrak{R}$ sei $\rho \leq h$; da $\mathfrak{U}'$ aus ν Exemplaren $\mathfrak{R}$ besteht, so ist die Ordnung von x auf $\mathfrak{U}'$ genau gleich $\rho\nu$. Ferner nimmt x auf $\mathfrak{U}'$ keinen Wert mehr als ρ-fach an. Die Zahl g von § 2 hat im vorliegenden Falle den Wert $\rho\nu$, ferner hat man für die Zahlen h und r diesmal die oberen Abschätzungen $hp^3\nu : n^2$ und ρ. Nach § 2 konvergiert dann eine Teilfolge der $\mathfrak{p}$ gegen einen auf $\mathfrak{U}'$ gelegenen Pol $\mathfrak{p}_0$ von x, und in diesen $\mathfrak{p}$ gilt für jede in $\mathfrak{p}_0$ verschwindende und auf $\mathfrak{U}'$ rationale Funktion $\phi(\mathfrak{p})$ die Abschätzung (91), nämlich

$$|\phi(\mathfrak{p})| < c_9 N(|\mu_k \Phi_1| + |\mu_k \Phi_k| + |\mu_k|)^{-\frac{n^2}{hlp^3} + \varepsilon}$$

für $k = 2, \cdots, p$. Durch Multiplikation dieser Ungleichungen folgt für $p > 1$ mit Rücksicht auf (113) und (114)

$$(115)\qquad |\phi(\mathfrak{p})| < c_{11} N(|\zeta_0| + \cdots + |\zeta_p|)^{-\varkappa n^2 + \varepsilon},$$

wo $\varkappa$ die von n unabhängige Zahl $1 : hlp^3(p-1)$ bedeutet, ε irgendeine positive Zahl ist und c_{11} nicht von $\mathfrak{p}$ abhängt.

Man kann noch voraussetzen, daß die Werte $\Phi_1(\mathfrak{p}_0), \cdots, \Phi_p(\mathfrak{p}_0)$ sämtlich endlich sind, da man sonst nur die Parameter $\alpha_1, \cdots, \gamma_2$ in (112) abzuändern brauchte. Setzt man $\Phi_k(\mathfrak{p}_0) = \omega_k$, so ist nach (115) speziell

$$(116)\qquad |\Phi_k(\mathfrak{p}) - \Phi_k(\mathfrak{p}_0)| = \left|\frac{\zeta_k}{\zeta_0} - \omega_k\right| < c_{11} N(|\zeta_0| + \cdots + |\zeta_p|)^{-\varkappa n^2 + \varepsilon}$$

für $k = 1, \cdots, p$. Hierauf kann man nun aber nicht ohne weiteres denselben Schluß anwenden wie in § 3; denn der Grad der algebraischen Zahl ω_k könnte die Größenordnung n^{2p} besitzen, da ja ω_k durch die n-Teilung der Perioden der Abelschen Funktionen bestimmt wird; und n^{2p} wächst für $p \geq 2$ nicht schwächer an als das Quadrat des Exponenten auf der rechten Seite von (116). In (116) sind jedoch p Approximationsaussagen enthalten, und die approximierten Zahlen ω_k sind, wie sogleich nach § 4 gezeigt werden soll, »hinreichend« unabhängig voneinander. Dieser Umstand ermöglicht es, die Methode der Einleitung erfolgreich durchzuführen und einen Widerspruch zu (116) abzuleiten.

Im folgenden braucht man nämlich den Hilfssatz:

Es sei δ kleiner als jede der beiden Zahlen $\varkappa n^2$ und $\frac{n^2}{hp^{2p+1}}$. Dann besteht zwischen den Zahlen $\omega_1, \cdots, \omega_p$ keine algebraische Gleichung δ-ten Grades mit Koeffizienten aus $\mathfrak{K}$. Anders ausgedrückt: Kein Polynom δ-ten Grades in $\Phi_1(\mathfrak{p}), \cdots, \Phi_p(\mathfrak{p})$ mit Koeffizienten aus $\mathfrak{K}$ hat die Nullstelle $\mathfrak{p} = \mathfrak{p}_0$.

Es sei $G(\mathfrak{p})$ ein Polynom vom Grade δ in $\Phi_1(\mathfrak{p}), \cdots, \Phi_p(\mathfrak{p})$ mit ganzen Koeffizienten aus $\mathfrak{K}$, das für $\mathfrak{p} = \mathfrak{p}_0$ verschwindet. Nach § 4 verschwindet es nicht identisch in $\mathfrak{p}$, denn δ ist ja kleiner als $\frac{n^2}{hp^{2p+1}}$. Nach (115) ist ferner für unendlich viele $\mathfrak{p} \to \mathfrak{p}_0$

$$(117)\qquad |G(\mathfrak{p})| < c_{11} N(|\zeta_0| + \cdots + |\zeta_p|)^{-\varkappa n^2 + \varepsilon}.$$

Nun hat $\zeta_0^\delta G(\mathfrak{p})$ eine Majorante der Form $c_{12}(|\zeta_0|+\cdots+|\zeta_p|)^\delta$; andererseits ist $\zeta_0^\delta G(\mathfrak{p})$ ganz und $\neq 0$, falls $\mathfrak{p}$ hinreichend dicht bei $\mathfrak{p}_0$ liegt; die Norm dieser Zahl ist also absolut genommen mindestens gleich 1. Dies liefert wegen (117) die Ungleichung

$$1 < c_{11} N(|\zeta_0|+\cdots+|\zeta_p|)^{-\varkappa n^2+1} |\zeta_0|^\delta c_{13} \frac{N(|\zeta_0|+\cdots+|\zeta_p|)^\delta}{(|\zeta_0|+\cdots+|\zeta_p|)^\delta},$$

also erst recht

$$N(|\zeta_0|+\cdots+|\zeta_p|)^{\varkappa n^2-1-\delta} < c_{14},$$

und dies ist wegen $\delta < \varkappa n^2$ ein Widerspruch. Damit ist der Hilfssatz bewiesen.

§ 6. Anwendung der Approximationsmethode.

Der durch Adjunktion der Zahlen $\omega_1, \cdots, \omega_p$ zu $\mathfrak{K}$ entstehende Körper $\mathfrak{K}'$ sei vom Grade d. Zur Bestimmung der Zahlen $\omega_1, \cdots, \omega_p$ hat man nur eine algebraische Gleichung n^{2p}-ten Grades mit Koeffizienten aus $\mathfrak{K}$ zu lösen, nämlich die Gleichung, von welcher die n-Teilung der Perioden der Abelschen Funktionen abhängt. Daher ist $d \leq l n^{2p}$. Ohne Beschränkung der Allgemeinheit kann man voraussetzen, daß ω_1 den Körper $\mathfrak{K}'$ erzeugt. Wäre dies nämlich nicht der Fall, so hätte man nur an die Stelle von $\Phi_1(\mathfrak{p})$ eine geeignete feste lineare Verbindung von $\Phi_1(\mathfrak{p}), \cdots, \Phi_p(\mathfrak{p})$ mit Koeffizienten aus $\mathfrak{K}$ zu setzen, und dabei blieben die Ungleichungen (116) ungeändert, abgesehen von dem Wert der Konstanten c_{11}. Ferner kann man annehmen, daß $\omega_1, \cdots, \omega_p$ ganze Zahlen sind, da man sonst nur $\Phi_1, \cdots, \Phi_p$ mit dem Hauptnenner dieser Zahlen multiplizieren müßte. Die Zahl ω_1 werde kurz mit ω bezeichnet.

Es sei λ die kleinere der beiden Zahlen $\varkappa = 1 : h l p^3 (p-1)$ und $1 : h p^{2p+1}$. Es sei

$$n > \lambda^{-\frac{1}{2}}$$

und

(118) $$1 \leq \delta < \lambda n^2.$$

Die Anzahl aller Potenzprodukte von $\omega_1, \cdots, \omega_p$, deren Dimension $\leq \delta$ ist, beträgt genau

(119) $$\binom{\delta+p}{p} = m+1;$$

sie seien in irgendeiner Reihenfolge mit $\alpha_0, \cdots, \alpha_m$ bezeichnet. Unter ihnen kommt speziell die Zahl 1 vor. Wegen (118) folgt aus dem Hilfssatz des vorigen Paragraphen, daß die Zahlen $\alpha_0, \cdots, \alpha_m$ im Körper $\mathfrak{K}$ linear unabhängig sind. Nun seien $P_0(x), \cdots, P_m(x)$ Polynome in x vom Grade q mit ganzen rationalen Koeffizienten. Damit das Polynom

$$P(x) = \alpha_0 P_0(x) + \cdots + \alpha_m P_m(x)$$

an der Stelle $x = \omega$ mindestens von der Ordnung b verschwindet, müssen die Gleichungen

(120) $$\alpha_0 P_0^{(k)}(\omega) + \cdots + \alpha_m P_m^{(k)}(\omega) = 0,$$

deren linke Seiten aus $P(x)$ durch k-malige Differentiation entstanden sind, für $k = 0, 1, \cdots, b-1$ erfüllt sein. Dies sind b homogene lineare Gleichungen für die $(m+1)(q+1)$ Koeffizienten von $P_0(x), \cdots, P_m(x)$. Drückt man alle in diese Gleichungen eingehenden Zahlen aus $\mathfrak{K}'$ durch eine Basis in bezug auf den Körper der rationalen Zahlen aus, so zerfällt jede einzelne Gleichung in d neue Gleichungen mit ganzen rationalen Koeffizienten; man hat also insgesamt bd Gleichungen mit $(m+1)(q+1)$ Unbekannten.

Für jedes natürliche a ist die Zahl $\binom{a}{k}$ ganz. Daher haben die ganzen rationalen Koeffizienten der d aus (120) entstandenen Gleichungen den gemeinsamen Teiler $k!$; durch diesen werde dividiert. Beachtet man, daß $\binom{a}{k} < 2^a$ ist, so ist ersichtlich, daß nunmehr alle ganzen rationalen Koeffizienten der bd Gleichungen absolut genommen kleiner als c_{15}^q sind. Dabei bedeutet c_{15} eine Zahl, die nicht von b und q abhängt; dieselbe Bedeutung sollen weiterhin $c_{16}, \cdots, c_{26}$ besitzen.

Nun sei

(121) $$b > 2(m+1)^2(2m+1)d.$$

Man lege q fest durch

(122) $$q = \left[\frac{b}{m+1}\left(d + \frac{1}{2m+1}\right)\right] - 1.$$

Setzt man noch

$$\vartheta = \frac{(m+1)(q+1)}{bd} - 1,$$

so ist nach (121) und (122)

(123) $$\frac{1}{(2m+2)d} < \vartheta \leq \frac{1}{(2m+1)d}.$$

Jetzt wende man den zweiten Hilfssatz der Einleitung an; die dort mit n und m bezeichneten Zahlen haben hier die Werte $(m+1)(q+1)$ und bd, ferner wird A durch c_{15}^q majorisiert. Es lassen sich also die Koeffizienten von $P_0(x), \cdots, P_m(x)$ als ganze rationale Zahlen bestimmen, die nicht sämtlich 0 sind, absolut genommen kleiner als $1 + \{(m+1)(q+1)c_{15}^q\}^{\frac{1}{\vartheta}}$ sind und den Gleichungen (120) für $k = 0, \cdots, b-1$ genügen. Mit Rücksicht auf (122) und (123) ist die gefundene Schranke

(124) $$1 + \{(m+1)(q+1)c_{15}^q\}^{\frac{1}{\vartheta}} < c_{16}^b.$$

Von den Polynomen $P_0(x), \cdots, P_m(x)$ können einige linear abhängig voneinander sein, also auch linear abhängig in bezug auf den Körper der rationalen Zahlen. Es sei $\mu + 1$ die Anzahl der linear unabhängigen unter ihnen; die Bezeichnung sei so gewählt, daß $P_0(x), \cdots, P_\mu(x)$ linear unabhängig sind. Dann ist

$$P(x) = \beta_0 P_0(x) + \cdots + \beta_\mu P_\mu(x),$$

wo $\beta_0, \cdots, \beta_\mu$ homogene lineare Verbindungen von $\alpha_0, \cdots, \alpha_m$ mit rationalen Koeffizienten bedeuten. Da $\alpha_0, \cdots, \alpha_m$ linear unabhängig in $\mathfrak{K}$ sind, so sind $\beta_0, \cdots, \beta_\mu$ sämtlich ungleich 0. Es sei

$$W(x) = |P_l^{(k)}(x)|,$$

wo der Zeilenindex k und der Spaltenindex l die Werte $0, \cdots, \mu$ durchlaufen, die Wronskische Determinante der Polynome $P_0, \cdots, P_\mu$; sie verschwindet nicht identisch in x und ist ein Polynom vom Grade $(\mu+1)q$ mit rationalen Koeffizienten. Es sei $W_{kl}(x)$ die Unterdeterminante von $P_k^{(l)}$ in W. Aus den $\mu + 1$ Gleichungen

$$P^{(k)}(x) = \beta_0 P_0^{(k)}(x) + \cdots + \beta_\mu P_\mu^{(k)}(x) \qquad (k = 0, \cdots, \mu)$$

folgt dann

(125) $$\beta_k W(x) = \sum_{l=0}^{\mu} W_{kl}(x) P^{(l)}(x), \qquad (k = 0, \cdots, \mu)$$

und zwar gilt dies identisch in $\beta_0, \cdots, \beta_\mu$. Nun verschwindet $P(x)$ für $x = \omega$ mindestens von der Ordnung b, also $P^{(l)}(x)$ mindestens von der Ordnung $b - l$. Da $\beta_k \neq 0$ ist, so verschwindet nach (125) das Polynom $W(x)$ für $x = \omega$ mindestens von der Ordnung $b - \mu$. Dies gilt für alle d Konjugierten von ω. Daher ist

(126) $$(\mu+1)q - d(b-\mu) = g \geq 0,$$

und $W(x)$ verschwindet für jeden beliebigen von den Konjugierten von ω verschiedenen Wert ξ höchstens von der Ordnung g. Aus (121), (122) und (126) folgt aber

$$0 \leq g < \frac{\mu+1}{m+1}bd + \frac{b}{2m+1} + md - bd < b\left(\frac{\mu-m}{m+1}d + \frac{1}{2m+1} + \frac{m}{2(m+1)^2(2m+1)}\right)$$
$$< \frac{b}{m+1}\{(\mu-m)d + 1\},$$

also $\mu = m$ und

$$(127) \qquad g < \frac{b}{m+1}.$$

Wegen $\mu = m$ sind alle Polynome $P_0(x), \cdots, P_m(x)$ voneinander linear unabhängig, man kann also die α_k durch die β_k ersetzen $(k = 0, \cdots, m)$.

Es sei ξ verschieden von den d zu ω konjugierten Zahlen. Die Wronskische Determinante $W(x)$ verschwinde für $x = \xi$ von der Ordnung $\gamma \leq g$. Dann ist $W^{(\gamma)}(\xi) \neq 0$. Andererseits ist nach (125) für $k = 0, \cdots, m$ die Zahl $\alpha_k W^{(\gamma)}(\xi)$ eine homogene lineare Kombination von $P(\xi), P'(\xi), \cdots, P^{(m+\gamma)}(\xi)$, und zwar gilt dies identisch in $\alpha_0, \cdots, \alpha_m$. Daher sind unter den $m+\gamma+1$ Ausdrücken $P(\xi), P'(\xi), \cdots, P^{(m+\gamma)}(\xi)$, wenn sie als homogene lineare Funktionen von $\alpha_0, \cdots, \alpha_m$ angesehen werden, $m+1$ voneinander linear unabhängige. Dies seien die Ausdrücke $P^{(k)}(\xi)$ für $k = k_0, \cdots, k_m$. Bei Division durch $k!$ mögen hieraus die Werte

$$(128) \qquad \begin{aligned} Q_0(\xi) &= P_{00}(\xi)\alpha_0 + \cdots + P_{0m}(\xi)\alpha_m \\ &\cdots\cdots\cdots\cdots\cdots \\ Q_m(\xi) &= P_{m0}(\xi)\alpha_0 + \cdots + P_{mm}(\xi)\alpha_m \end{aligned}$$

hervorgehen. Die Determinante $|P_{kl}(\xi)|$ ist von 0 verschieden. Die $P_{kl}(\xi)$ sind Polynome g-ten Grades in ξ; ihre Koeffizienten sind ganz rational und nach (124) absolut kleiner als c_{17}^b. Ferner verschwinden $Q_0(x), \cdots, Q_m(x)$ für $x = \omega$ mindestens von der Ordnung $b' = b - m - \gamma$. Nach (121) und (127) ist

$$(129) \qquad b' \geq b - m - g > b\left(1 - \frac{m}{2(m+1)^2(2m+1)} - \frac{1}{m+1}\right) > \frac{b}{4}.$$

Aus der Formel

$$Q_k(\xi) = \frac{1}{2\pi i}(\xi - \omega)^{b'} \int \frac{Q_k(t)}{(t-\omega)^{b'}} \frac{dt}{t - \xi},$$

wo das Integral etwa über den Kreis $|t - \xi| = 1$ erstreckt wird, entnimmt man die Abschätzung

$$(130) \qquad |Q_k(\xi)| < c_{18}^b |\xi - \omega|^{b'} (1 + |\xi|)^{g-b'}. \qquad (k = 0, \cdots, m)$$

Aus den unendlich vielen Punkten $\mathfrak{p}$ mit dem Grenzwert $\mathfrak{p}_0$, für welche (116) gilt, mögen jetzt zwei herausgegriffen werden, etwa $\mathfrak{p}$ und $\mathfrak{p}'$. Die Werte $\Phi_k(\mathfrak{p}')$ seien mit $\zeta'_k : \zeta'_0$ bezeichnet, für $k = 1, \cdots, p$, so daß auch $\zeta'_0, \cdots, \zeta'_p$ teilerfremd sind; und es möge noch

$$(131) \qquad \log N(|\zeta_0| + \cdots + |\zeta_p|) = \Lambda, \qquad \log N(|\zeta'_0| + \cdots + |\zeta'_p|) = \Lambda'$$

gesetzt werden. Liegt $\mathfrak{p}'$ hinreichend nahe bei $\mathfrak{p}_0$, so ist $\zeta'_1 : \zeta'_0$ von ω und seinen Konjugierten verschieden; wäre nämlich $\zeta'_1 : \zeta'_0 = \omega$, so wäre ω eine Zahl aus $\mathfrak{K}$, was nach (118) zu § 5 in Widerspruch steht; und für hinreichend nahe bei $\mathfrak{p}_0$ gelegenes $\mathfrak{p}'$ kann $\Phi_1(\mathfrak{p}')$ auch nicht gleich einer von $\omega = \Phi_1(\mathfrak{p}_0)$ verschiedenen Konjugierten von ω sein. Also kann $\xi = \zeta'_1 : \zeta'_0$ gewählt werden. Nach (116), (130) und (131) ist

$$\left| Q_k\left(\frac{\zeta'_1}{\zeta'_0}\right) \right| < c_{19}^b e^{-\Lambda'(\varkappa n^2 - 1)b'}, \qquad (k = 0, \cdots, m)$$

also nach (129)

$$(132) \qquad \left| Q_k\left(\frac{\zeta'_1}{\zeta'_0}\right) \right| < c_{19}^b e^{-\varkappa_1 b \Lambda' n^2},$$

wo $\varkappa_1$ eine positive Konstante bedeutet, die nicht von b und n abhängt.

In den rechten Seiten von (128) ersetze man die Potenzprodukte $\alpha_0, \cdots, \alpha_m$ von $\omega_1, \cdots, \omega_p$ durch die analog gebildeten Potenzprodukte $A_0, \cdots, A_m$ von $\zeta_1 : \zeta_0, \cdots, \zeta_p : \zeta_0$. Dadurch treten an die Stelle von $Q_0(\xi), \cdots, Q_m(\xi)$ die Zahlen

$$Z_0 = P_{00}(\xi)A_0 + \cdots + P_{0m}(\xi)A_m$$

$$\cdots\cdots\cdots\cdots\cdots\cdots$$

$$Z_m = P_{m0}(\xi)A_0 + \cdots + P_{mm}(\xi)A_m$$

mit $\xi = \zeta_1' : \zeta_0'$. Diese Zahlen liegen sämtlich in $\mathfrak{K}$. Die Determinante $|P_{kl}(\xi)|$ ist $\neq 0$ und unter den Zahlen $A_0, \cdots, A_m$ kommt der Wert 1 vor. Daher ist mindestens eine der Zahlen $Z_0, \cdots, Z_m$ von 0 verschieden, etwa die Zahl $Z_0 = Z$. Nun ist jedes $P_{kl}(\xi)$ ein Polynom q-ten Grades in ξ, ferner sind die Potenzprodukte $A_0, \cdots, A_m$ höchstens vom Grade δ. Folglich ist die Zahl

$$(\zeta_0')^q \zeta_0^\delta Z = \Gamma$$

ganz. Die Konjugierten von Γ werden durch die Konjugierten von

$$c_{20}^b(|\zeta_0'| + \cdots + |\zeta_p'|)^q(|\zeta_0| + \cdots + |\zeta_p|)^\delta$$

majorisiert und Γ selbst durch

$$|Z|(|\zeta_0'| + \cdots + |\zeta_p'|)^q(|\zeta_0| + \cdots + |\zeta_p|)^\delta .$$

Daher ist mit (131)

(133) $$1 \leq |N\Gamma| < c_{21}^b e^{\Lambda' q + \Lambda\delta}|Z| .$$

Man benutze nun die Abschätzung

$$\left|\left(\frac{\zeta_1}{\zeta_0}\right)^{l_1} \cdots \left(\frac{\zeta_p}{\zeta_0}\right)^{l_p} - \omega_1^{l_1} \cdots \omega_p^{l_p}\right| < c_{22}\left(\left|\frac{\zeta_1}{\zeta_0} - \omega_1\right| + \cdots + \left|\frac{\zeta_p}{\zeta_0} - \omega_p\right|\right)$$

und nochmals (116); es folgt

$$\left|Z_k - Q_k\left(\frac{\zeta_1'}{\zeta_0'}\right)\right| < c_{23}^b e^{-\Lambda(\varkappa n^2 - 1)} \qquad (k = 0, \cdots, m)$$

und in Verbindung mit (132)

(134) $$|Z| < c_{24}^b(e^{-\varkappa_1 n^2 \Lambda} + e^{-\varkappa_1 n^2 b\Lambda'}) .$$

Die bisher noch willkürliche Zahl b werde durch

$$b = \left[\frac{\Lambda}{\Lambda'}\right]$$

festgelegt; da b der Ungleichung (121) genügen soll, so bedeutet dies eine Einschränkung für die Wahl von $\mathfrak{p}$, der man aber genügen kann, da Λ für $\mathfrak{p} \to \mathfrak{p}_0$ über alle Grenzen wächst. Dann ist $b\Lambda' \leq \Lambda < (b+1)\Lambda'$, und aus (133) und (134) ergibt sich

(135) $$\Lambda'(q + (b+1)\delta - \varkappa_1 n^2 b) + c_{25} b > 0 .$$

Für δ wähle man den Wert

(136) $$\delta = \left[n^{\frac{2p}{p+1}}\right];$$

damit (118) erfüllt ist, muß n so groß sein, daß

(137) $$\left[n^{\frac{2p}{p+1}}\right] < \lambda n^2$$

ist. Wegen $d \leq l n^{2p}$ ist nach (119), (122) und (136)

$$q < \varkappa_2 b n^{2p} : \left(n^{\frac{2p}{p+1}}\right)^p = \varkappa_2 n^{\frac{2p}{p+1}} b ,$$

wo $\varkappa_2$ nicht von n und b abhängt, ferner

$$(b+1)\delta < \varkappa_3 n^{\frac{2p}{p+1}} b .$$

mit analoger Bedeutung von $\varkappa_3$. Der Faktor von Λ' in (135) ist also kleiner als $b\{(\varkappa_2+\varkappa_3)n^{\frac{2p}{p+1}}-\varkappa_1 n^2\} = -c_{26}b$.

Nun wähle man n so groß, daß (137) erfüllt ist und daß die Zahl

$$c_{26} = n^2\left\{\varkappa_1-(\varkappa_2+\varkappa_3)n^{-\frac{2}{p+1}}\right\}$$

positiv ist, und darauf $\mathfrak{p}'$ derart, daß $\Lambda' > c_{25}:c_{26}$ ist. Dann entsteht ein Widerspruch zu (135).

Damit ist der Satz auch im Falle $p>1$ bewiesen.

§ 7. Kubische Formen mit positiver Diskriminante.

Die Untersuchungen der vorangehenden Paragraphen geben die Möglichkeit, eine Schranke für die Anzahl der Lösungen der diophantischen Gleichung $f(x, y) = 0$ als Funktion der Koeffizienten von f explicit aufzustellen, falls diese Gleichung nur endlich viele Lösungen besitzt. Man kann nun vermuten, daß sich sogar eine Schranke finden läßt, die nur von der Anzahl der Koeffizienten abhängt; doch dürfte dies recht schwer zu beweisen sein. Eine Stütze für diese Vermutung bilden die im folgenden entwickelten, allerdings sehr speziellen Resultate.

Durch die Untersuchungen von B. Delaunay und Nagell ist festgestellt, daß eine kubische Form mit ganzen rationalen Koeffizienten und negativer Diskriminante für ganze rationale Werte der Variabeln x, y den Wert 1 höchstens fünfmal annimmt. Die Beweise der beiden Autoren bedienen sich der Einheitentheorie und lassen sich, wie es scheint, nicht auf kubische Formen mit positiver Diskriminante und andere Werte als 1 oder 3 anwenden. Dies wird aber ermöglicht durch den Ansatz, den Thue zu Beginn seiner Untersuchungen über diophantische Gleichungen machte, nämlich die Verwendung des Kettenbruchs für $(1-x)^\alpha$; man hat nur Thues Abschätzungen ein wenig zu verfeinern.

Es seien m und n nicht negative ganze rationale Zahlen und F die hypergeometrische Funktion. Dann ist

$$(1-z)^\alpha F(-n+\alpha, -m, -m-n, z) - F(-m-\alpha, -n, -m-n, z)$$
$$= \lambda z^{m+n+1} F(n-\alpha+1, m+1, m+n+2, z)$$

mit

$$\lambda = (-1)^{n-1}\binom{m+\alpha}{m+n+1}:\binom{m+n}{n}.$$

Dies benutze man speziell für $m=n-g$ und $g=0, 1$. Man setze

$$A_g(z) = F(-n+\alpha, -n+g, -2n+g, z)$$
$$B_g(z) = F(-n-\alpha+g, -n, -2n+g, z)$$
$$R_g(z) = \lambda_g z^{2n-g+1} F(n-\alpha+1, n-g+1, 2n-g+2, z)$$
$$\lambda_g = (-1)^{n-g}\binom{n+\alpha-g}{2n-g+1}:\binom{2n-g}{n},$$

dann ist

$$(1-z)^\alpha A_g(z) - B_g(z) = R_g(z); \qquad (g=0, 1)$$

und da $A_g(z)$ ein Polynom vom Grade $n-g$, $B_g(z)$ ein Polynom vom Grade n ist, so ist der Ausdruck

$$(138) \qquad A_0B_1 - A_1B_0 = -A_0R_1 + A_1R_0 = -\lambda_1 z^{2n} + \cdots = -\lambda_1 z^{2n},$$

also $\neq 0$ für $z\neq 0$ und $\alpha \neq -n+1, -n+2, \ldots, n$.

Nach Riemann ist für $n\to\infty$

$$B_g \sim (1-z)^{\frac{\alpha}{2}-\frac{1}{4}}\left(\frac{\sqrt{1-z}+1}{2}\right)^{2n+1-g}$$

$$R_g \sim -2 \sin \pi \alpha (1-z)^{\frac{\alpha}{2}-\frac{1}{4}} \left(\frac{\sqrt{1-z}-1}{2}\right)^{2n+1-g},$$

und zwar gleichmäßig in jedem abgeschlossenen Gebiet der z-Ebene, das die Punkte 1 und ∞ nicht enthält. Also ist auch für beliebiges z'

$$(139) \quad (1-z)^n A_g - (1-z')^n B_g = R_g + (1-(1-z')^\alpha) B_g$$

$$\sim (1-z)^{\frac{\alpha}{2}-\frac{1}{4}} \left\{ -2 \sin \pi \alpha \left(\frac{\sqrt{1-z}-1}{2}\right)^{2n+1-g} + (1-(1-z')^\alpha)\left(\frac{\sqrt{1-z}+1}{2}\right)^{2n+1-g} \right\}.$$

Hierin ist die linke Seite nach (138) entweder für $g=0$ oder für $g=1$ von 0 verschieden, wenn noch vorausgesetzt wird, daß z von 0 und 1 verschieden und α keine ganze rationale Zahl ist. Es sei fortan $\alpha = \frac{1}{3}$.

Nach dem Satze 5 des ersten Teiles dieser Abhandlung ist der Hauptnenner h der Koeffizienten von A_0, A_1, B_0, B_1 höchstens gleich γ_1^n, wo γ_1 eine positive Konstante bedeutet.

Es sei

$$\phi(x, y) = a_0 x^3 + a_1 x^2 y + a_2 x y^2 + a_3 y^3$$

eine kubische Form mit ganzen rationalen Koeffizienten a_0, a_1, a_2, a_3 und positiver Diskriminante

$$d = a_1^2 a_2^2 - 4 a_0 a_2^3 - 4 a_1^3 a_3 - 27 a_0^2 a_3^2 + 18 a_0 a_1 a_2 a_3.$$

Zwischen den Kovarianten

$$\psi(x, y) = -\frac{1}{4}\left(\frac{\partial^2 \phi}{\partial x^2}\frac{\partial^2 \phi}{\partial y^2} - \frac{\partial^2 \phi}{\partial x \partial y}\frac{\partial^2 \phi}{\partial x \partial y}\right)$$

$$\chi(x, y) = \frac{\partial \phi}{\partial x}\frac{\partial \psi}{\partial y} - \frac{\partial \phi}{\partial y}\frac{\partial \psi}{\partial x}$$

und ϕ besteht die Identität

$$(140) \qquad 4\psi^3 = \chi^2 + 27 d \phi^2.$$

Die Form $\psi(x, y)$ ist quadratisch und hat die negative Diskriminante $-3d$. Aus (140) folgt in bekannter Weise

$$(141) \qquad \psi = \xi\eta, \quad \chi = \xi^3 + \eta^3, \quad \phi = \frac{\xi^3 - \eta^3}{3\sqrt{-3d}},$$

$$(142) \qquad \xi^3 = \tfrac{1}{2}(\chi + 3\phi\sqrt{-3d}), \quad \eta^3 = \tfrac{1}{2}(\chi - 3\phi\sqrt{-3d}),$$

wo ξ und η lineare Funktionen von x und y bedeuten. Es gibt zwei Konstanten λ und μ, so daß $\xi:\lambda$ und $\eta:\mu$ lineare Funktionen von x und y mit Koeffizienten aus dem durch $\sqrt{-3d}$ erzeugten imaginär quadratischen Körper $\mathfrak{K}$ sind. Sind x und y ganz rational, so sind es auch ϕ, ψ, χ; nach (141) sind dann ξ^3 und η^3 konjugierte ganze Zahlen aus $\mathfrak{K}$; und die Zahlen $\xi:\lambda$ und $\eta:\lambda$ liegen ebenfalls in $\mathfrak{K}$.

Nun sei vorgelegt die diophantische Gleichung

$$(143) \qquad \phi(x, y) = k,$$

wo k eine feste natürliche Zahl bedeutet. Für jede Lösung x, y dieser Gleichung bilde man die durch (142) definierten Zahlen ξ^3 und η^3 und setze noch

$$\frac{3k\sqrt{-3d}}{\xi^3} = z.$$

Dann ist nach (141) und (143)

$$\left(\frac{\eta}{\xi}\right)^3 = 1 - z,$$

also

(144) $$\frac{\eta}{\xi} = \varepsilon(1-z)^{\frac{1}{3}},$$

wo $(1-z)^{\frac{1}{3}}$ den Hauptwert hat und ε eine dritte Einheitswurzel bedeutet. Nun sei x', y' eine zweite Lösung von (143), für welche die Einheitswurzel ε in (144) denselben Wert hat, und es seien ξ', η', z' die zu x', y' gehörigen Werte von ξ, η, z. Bezeichnet man die Zahl

$$(1-z)^{\frac{1}{3}} A_g(z) - (1-z')^{\frac{1}{3}} B_g(z) = \varepsilon^{-1}\left(\frac{\eta}{\xi} A_g(z) - \frac{\eta'}{\xi'} B_g(z)\right)$$

mit δ, so ist δ^3 eine Zahl von $\mathfrak{K}$, weil $\left(\frac{\eta}{\xi}\right)^3$ und die Quotienten $\xi:\xi'$, $\eta:\eta'$ es sind. Ferner ist die Zahl $h\xi'\xi^{3n+1-g}\delta$ ganz; ist sie $\neq 0$, so ist ihr absoluter Betrag also mindestens gleich 1. Man wähle nun $g = 0$ oder $= 1$ derart, daß $\delta \neq 0$ ist; dann ist

(145) $$|\delta| \geq \frac{1}{h|\xi'\xi^{3n+1-g}|} = \frac{|z'|^{\frac{1}{3}}|z|^{n+\frac{1-g}{3}}}{h|3k\sqrt{-3d}|^{n+\frac{2-g}{3}}}.$$

Da $|1-z| = 1$ ist, also z weder in der Nähe von 1 noch in der Nähe von ∞ liegt, so läßt sich (139) anwenden. Mit Rücksicht auf die Ungleichung $h \leq \gamma_1^n$ und (145) folgt

(146) $$|z|^{\frac{1}{3}}|z|^{n+\frac{1-g}{3}} < \gamma_2^n (k\sqrt{d})^{n+\frac{2-g}{3}}(|z|^{2n+1-g} + |z'|)$$

mit konstantem γ_2.

Sind ε_1, ε_2, ε_3 die drei dritten Einheitswurzeln, so gilt

$$(\varepsilon_1\xi-\eta)(\varepsilon_2\xi-\eta)(\varepsilon_3\xi-\eta) = 3k\sqrt{-3d}.$$

Von den drei Faktoren der linken Seite sei etwa $\varepsilon_1\xi-\eta$ absolut genommen am kleinsten. Da $\eta:\xi$ vom absoluten Betrage 1 ist, so wird $|\varepsilon_2\xi-\eta| \geq |\xi|$, $|\varepsilon_3\xi-\eta| \geq |\xi|$, und daher

$$|\varepsilon_1\xi-\eta| \leq \frac{|3k\sqrt{-3d}|}{|\xi|^2}.$$

Ist nun $|z| < 1$, so ist erst recht $\left|1-(1-z)^{\frac{1}{3}}\right| < 1$, also nach (144) auch $|\varepsilon\xi-\eta| < |\xi|$, und folglich $\varepsilon_1 = \varepsilon$. Ist aber $|z| \geq 1$, so gilt jedenfalls wegen $|\eta| = |\xi|$ die Ungleichung

$$|\varepsilon\xi-\eta| \leq 2|\xi| \leq 2|\xi z| = \frac{|6k\sqrt{-3d}|}{|\xi|^2}.$$

Daher ist stets

$$|\varepsilon\xi-\eta| \leq \frac{|6k\sqrt{-3d}|}{|\xi|^2}$$

und auch

$$|\varepsilon\xi'-\eta'| \leq \frac{|6k\sqrt{-3d}|}{|\xi'|^2},$$

also

(147) $$|\xi\eta'-\eta\xi'| \leq 6k|\sqrt{-3d}|\left(\frac{|\xi'|}{|\xi|^2} + \frac{|\xi|}{|\xi'|^2}\right).$$

Die quadratische Form $\xi\eta$ hat die Diskriminante $-3d$; dies liefert

$$\xi\eta'-\eta\xi' = (xy'-yx')\sqrt{-3d}.$$

Sind nun die beiden Lösungen x, y und x', y' von (143) nicht identisch, so ist $xy'-yx' \neq 0$, und daher $|\xi\eta'-\eta\xi'| \geq |\sqrt{-3d}|$. Ist $|\xi'| \geq |\xi|$, so folgt aus (147)

$$|\xi|^2 \leq 12k|\xi'|$$

$$|z|^2 = \frac{|3k\sqrt{-3d}|^2}{|\xi|^6} \geq \frac{|3k\sqrt{-3d}|^2}{(12k)^3|\xi'|^3} = \frac{|3k\sqrt{-3d}|}{(12k)^3}|z'|$$

(148) $$\left|\frac{\gamma_3 k^2}{\sqrt{d}} z\right|^2 \geq \left|\frac{\gamma_3 k^2}{\sqrt{d}} z'\right|.$$

Man ordne nun die sämtlichen Lösungen von (143), für welche die Einheitswurzel ε in (144) einen und denselben Wert hat, nach steigenden Werten von $|\xi|$, etwa $|\xi_0| \leq |\xi_1| \leq |\xi_2| \leq \cdots$. Für die zugehörigen Werte $z_0, z_1, z_2, \cdots$ von z gilt dann nach (148)

$$\left|\frac{\gamma_3 k^2}{\sqrt{d}} z_{r+s}\right| \leq \left|\frac{\gamma_3 k^2}{\sqrt{d}} z_r\right|^{2^s}. \qquad (r = 0, 1, \cdots; s = 1, 2, \cdots)$$

Existieren mindestens sieben zu ε gehörige Lösungen, so wähle man speziell $z = z_3$, $z' = z_6$. Wegen $|z_0| \leq 2$ ist dann

(149) $$|z| < \gamma_4 \left(\frac{k^2}{\sqrt{d}}\right)^7.$$

Setzt man noch

(150) $$d \geq \gamma_3^2 k^4$$

voraus, so ist ferner in der Gleichung

$$|z'| = |z|^\nu$$

der Exponent $\nu \geq 8$. Nach (146) wird

$$\gamma_2^n (k\sqrt{d})^{n+\frac{2-g}{3}} \left(|z|^{n+\frac{2}{3}(1-g)-\frac{\nu}{3}} + |z|^{\frac{2}{3}\nu-n-\frac{1}{3}(1-g)}\right) > 1.$$

Hierin setze man $n = \left[\frac{\nu+g}{2}\right]$, dann ist

$$n + \frac{2}{3}(1-g) - \frac{1}{3}\nu \geq \frac{1}{6}\nu - \frac{1}{2} > 0$$

$$\frac{2}{3}\nu - n - \frac{1}{3}(1-g) \geq \frac{1}{6}\nu - \frac{1}{2} > 0$$

$$n + \frac{1}{3}(2-g) \leq \frac{1}{2}\nu + \frac{5}{6}.$$

Dies liefert mit Rücksicht auf (149) die Abschätzung

$$\gamma_5^\nu (k\sqrt{d})^{\frac{1}{2}\nu+\frac{5}{6}} \left(\frac{k^2}{\sqrt{d}}\right)^{\frac{7}{6}\nu-\frac{7}{2}} > 1,$$

also erst recht

$$\gamma_6 (k\sqrt{d})^{\frac{1}{2}\cdot 8+\frac{5}{6}} \left(\frac{k^2}{\sqrt{d}}\right)^{\frac{7}{6}\cdot 8-\frac{7}{2}} > 1.$$

Diese Ungleichung enthält einen Widerspruch, falls

(151) $$d > \gamma_7 k^{33}$$

ist. Ist d so groß, daß (150) und (151) erfüllt sind, so existieren also höchstens sechs zu ε gehörige Lösungen und insgesamt höchstens 18 Lösungen von $\phi(x, y) = k$. Diese

Schranke 18 läßt sich übrigens durch einige Kunstgriffe noch weiter herabdrücken, doch soll darauf nicht mehr eingegangen werden.

Für die endlich vielen positiven Diskriminanten, die nicht (150) und (151) genügen, existieren nur endlich viele Klassen nicht äquivalenter kubischer Formen. Damit ist bewiesen, daß die Anzahl der Lösungen von $\phi(x, y) = k$ für beliebiges d unterhalb einer nur von k abhängigen Schranke liegt.

Dieselbe Überlegung läßt sich auch auf kubische Formen mit negativer Diskriminante d anwenden, doch erhält man für kleine Werte von $|d|$ nicht die scharfen Resultate von Delaunay und Nagell.

Noch einfacher ist die Untersuchung der diophantischen Gleichung

$$ax^n - by^n = k$$

für beliebiges festes $n \geq 3$. Es ergibt sich, daß sie höchstens eine Lösung in natürlichen Zahlen x, y besitzt, falls $|ab|$ eine nur von k und n abhängige Schranke übersteigt. Für $n = 3$ und $k = 1$ oder 3 ist diese Aussage in einem präziseren Satze von Nagell enthalten.

Integral points on curves: Siegel's theorem after Siegel's proof

Clemens Fuchs and Umberto Zannier

1 Introduction

In this article, conceived as an *addendum of comments* to the present translation of Siegel's paper, we shall present (in brief form) some fairly modern arguments for Siegel's theorem on integral points on curves, appearing in the second part of his paper [48] that is translated here; we shall refer to the more modern statement appearing as Theorem 3.2 below. The arguments presented here appeared after the original proof. All of these proofs rely on Diophantine Approximation and use suitable versions of Roth's theorem (1955) or Schmidt's Subspace Theorem (about 1970). Siegel had not Roth's theorem [45], which led to considerable complications in his proof. Some of the arguments below may be considered versions of Siegel's one, simplified both by the use of Roth's theorem and also by geometrical results on abelian varieties.[1]

Before this, we are going to recall, in short, some of the main developments which occurred in the topic, after Siegel's paper.

2 Some developments after Siegel's proof

In this section we comment on some developments that took place after Siegel published his paper. We do not claim any completeness of our references; we are just presenting a selection of results which to our taste are particularly relevant in comparison with Siegel's theorem.

2.1 Integral and rational points

Integral points in classical language are implicitly understood to be related to *affine*, rather than *projective*, varieties. However one may conceive integral points for a projective (or complete) variety as being just rational points. In 1982 G. Faltings [29] proved results which implied as a corollary the celebrated Mordell's conjecture predicting the finiteness of the set of *k-rational* points, over any number field k, on any complex irreducible curve of genus at least 2.

[1] This article is modelled on the second author's survey paper [59] and also [56].

Of course this statement (tauto)logically 'contains' Siegel's theorem, at any rate for genus 2 or more; but in fact simple arguments (based in practice on the Chevalley-Weil theorem) allow to recover from Faltings' theorem even the cases of genus 0 or 1 of Siegel's theorem. In spite of this major advance by Faltings, Siegel's theorem retains, however, interest and importance, and its proof is distinctly less demanding than Faltings'.

One should also recall that a more 'elementary' proof of Faltings' theorem was given by E. Bombieri [9], after another proof by P. Vojta [54]; Bombieri's approach related to classical Diophantine approximation, and enlightened common principles of this with Vojta's ideas (which in turn extended a method of D. Mumford). See also [11] for an exposition of this and much of the above.

2.2 Extensions to S-integers

In the 1930s, K. Mahler started to develop Diophantine approximation over p-adic fields; this led eventually to a generalisation of Siegel's theorem to 'S-integral points' (*cf.* [41]), namely points whose coordinates are rational numbers whose denominators are divisible only by primes in the finite set S. (Moreover, one may work with any number field in place of $\mathbb{Q}$.) Mahler also dealt with some cases of the theorem over function fields (a terminology not of completely common use at that time).

S. Lang, for instance in his books on Diophantine geometry (see [37]), further extended Mahler's context, and formulated a 'Roth's theorem' for an arbitrary finite set of places of a number field.[2] Lang also used this to give a proof of Siegel's theorem over arbitrary finitely generated rings (hence including rings of S-integers in number fields). Lang's proof may be considered in several respects a version of Siegel's; however the logic of these proofs in our opinion bears some differences, even if both proofs rely at bottom on the same geometrical properties of abelian varieties. (A version of Lang's proof shall be given in the first of the arguments below; some of Siegel's principles however appear in closer form in our second proof below. See also J-P. Serre's exposition in [46], where the stronger conclusion is reached that there are only finitely many rational points having a denominator 'sensibly' smaller than the numerator. See further [35].)

In the case of the affine curve $\mathbb{P}_1 \setminus \{0, 1, \infty\}$ and the ground field $\mathbb{Q}$, (Mahler's sharpening of) Siegel's theorem amounts to the finiteness of the solutions to $x + y = 1$ in S-units x, y (that is, x, y are rationals whose numerator and denominator are divisible only by primes from the finite set S).[3] This special case, though dealing only with the curve $\mathbb{P}_1$, is already deep. (By the way, already

[2] This shall be recalled below. Lang went even beyond that, axiomatising a theory of Diophantine approximation over fields 'with a product formula'.

[3] This was actually first proved by Mahler, whereas Siegel had considered this equation in number fields, but only for classical *units*, *i.e.*, units at all places. See below for a quick deduction of the general case from Lang's form of Roth's theorem.

Siegel had shown how to derive from this kind of result finiteness for integral points on certain curves of higher genus.) More recently, a quite sophisticated proof of this result has been found by M. Kim [36], by methods completely different from Siegel's and Faltings'. (See also Faltings' exposition [33].) However, to date such a method has not yet been extended to cover sensibly more general cases of Siegel's theorem. A further sophisticated method is in the paper by H. Pasten and M. Ram Murty [43], which uses results on modular forms, through an approach related to G. Frey's ideas concerning Fermat's Last Theorem.

2.3 Effectivity

Siegel's theorem was not, and is not yet, an effective result, in the sense that it does not allow to find the actual integral solutions, for a given (though arbitrary) equation. (The ineffectivity comes from the Diophantine approximation.) However the proof allows to bound effectively the number of solutions; on this last aspect and related ones, much significant work has been done after Siegel's time, notably by E. Bombieri, W.M. Schmidt, J.-H. Evertse and H.P. Schlickewei, actually in the general context of Diophantine approximation. (See *e.g.* [21,26–28,51] and Bombieri and W. Gubler's book [11] for an account of some of these works. See also [14] for an application to Siegel's theorem of bounds by Evertse in Diophantine approximation. Furthermore, for a collection of other references see [14,20,22,23,42,49,50].)

Around 1970, A. Baker proved certain lower bounds for non vanishing linear forms in logarithms of algebraic numbers, with algebraic coefficients. This allowed to prove effective bounds for the integral (or S-integral) points on certain classes of affine curves; these include curves defined by Thue's equations (*cf.* [12]), and also arbitrary curves of genus 1.[4] For these equations and curves, an effective method alternative to Baker's was found by Bombieri; see [10]. (For a recent and completely different approach, working over $\mathbb{Q}$, see also the already quoted paper [43].)

Other instances of effectivity for Siegel's theorem (derived again from Baker's results) occur for example when the curve is a Galois – ramified – cover of the affine line (see Y. Bilu's papers [6, 7], also for more general criteria, and the second author's paper [55]).[5]

However, already for genus 2 there are cases which escape from any of the known methods.

On the other hand, the theorem is known to be fully effective over function fields; actually, in this context even Faltings' theorem on rational points (see above) is known in effective versions, with a number of different proofs (in part obtained before Faltings'), on which we do not pause here.

[4] See for instance [4] or [5] for a more recent result.

[5] Concerning effectivity in Siegel's theorem, see also Coates' letter to Cassels reproduced as an appendix to [59].

2.4 Higher dimensions

As to integral points on (affine) algebraic varieties of higher dimension, the picture is less complete, in spite of a number of deep theorems, part of which we now recall (see also [11,38] and the survey paper of the second author [57]).

A remarkable result was obtained by W.M. Schmidt around 1970: by means of his Subspace Theorem[6] he succeeded to generalise the results on Thue's equation to *norm form equations* in any number of variables (see for instance [52,53]).[7] This was later extended by P. Corvaja and the second author in [15] and further by J.-H. Evertse and R. Ferretti [25], again using the Subspace Theorem. In the meantime J.-H. Evertse, and independently A. van der Poorten and H.P. Schlickewei, had generalised the S-unit equation to several variables, and later M. Laurent succeeded to use this and prove (the toric case of) a conjecture of Lang, describing integral points on subvarieties of $\mathbb{G}_{\mathrm{m}}^n$: their Zariski closure is found to be a finite union of translates of algebraic subgroups (see [11] for an account of these results and *e.g.* [10]).

In the context of abelian varieties, it is a difficult theorem of Faltings [30] that the complement of an abelian variety with respect to an ample divisor contains only finitely many integral points; P. Vojta extended this to semiabelian varieties. Faltings, relying in part on ideas due to Vojta, also proved in [31] Lang's conjecture on rational points on subvarieties of abelian varieties: their Zariski closure is a finite union of translates of abelian subvarieties. This result contains the Mordell conjecture, proved earlier by Faltings by other methods.

These achievements represent (remarkable) special cases of very general conjectures of Vojta, predicting that the integral (or rational) points on a variety should be *degenerate*, *i.e.* not Zariski-dense, provided suitable purely geometrical conditions involving the canonical class and the divisor at infinity are verified. (See [11] and [38].)

In this direction, for varieties which cannot be embedded (or mapped nontrivially) in semiabelian varieties, the known results are much more fragmentary. Still relying on Schmidt's Subspace Theorem, and using suitable projective embeddings, the paper [16] proves degeneracy of integral points on surfaces whose divisor at infinity satisfies certain purely numerical conditions of intersection[8] (see also [17] for some special applications obtained using the main result of [16]); A. Levin [39] and P. Autissier [3] independently deduced from this result that a surface whose divisor at infinity contains at least *four* distinct ample components cannot have a Zariski-dense set of integral points.[9] They

[6]This is a higher-dimensional extension of Roth's theorem, recalled in our third argument below.

[7]These equations are special cases of equations of the shape $L_1 \cdots L_n = 1$, where L_i are linear forms over $\overline{\mathbb{Q}}$ in $r \leq n-1$ variables, the case $r = 2$ being Thue's.

[8]See the third proof below for the principle of this, applied to the situation of Siegel's theorem.

[9]This should be compared to the *three* missing points appearing in Siegel's theorem. Indeed, it is not too difficult to deduce directly Siegel's theorem from any of these results.

also proved by similar methods some results in arbitrary dimensions. However, already for the affine surfaces obtained as the complement in $\mathbb{P}_2$ of three or less curves, very little is known; another paper of Faltings [32] succeeds to deal with the complement of a single curve in $\mathbb{P}_2$, but it is required that this curve is the branching divisor of a map from a surface, satisfying suitable numerical conditions. (See [16] for a complete treatment of the function field case of $\mathbb{P}_2$ minus three components.) For a newer result see also [19]. Furthermore we mention [58] for other results which use the method of embedding a variety in projective spaces of high dimension, in order to get information on the set of its integral points.[10]

The general results of Faltings and Vojta have also allowed to study rational or integral points on a curve, this time allowing them to be defined over a *variable* number field of bounded degree. For rational points, see [24] for an account; for integral points (on affine curves) the conclusions are more complete, due to a recent result by Levin relying also on an application of the Faltings-Vojta theorems obtained by J. Noguchi and J. Winkelmann.

See, further, Levin's paper [40].

Again, for function fields more is known; we just mention the paper [18].

2.5 Analogies with complex geometry

Broadening the context, it had been noted by C. Osgood, E. Reyssat and especially Vojta how Siegel's theorem bears an impressive and surprising analogy with certain other (sometimes celebrated) results in complex analysis and geometry. For instance, the finiteness of solutions of the above mentioned S-unit equation $x + y = 1$ may be interpreted in the holomorphic context as stating that the equation has no non-constant solutions in invertible entire functions $x = x(z)$, $y = y(z)$ on $\mathbb{C}$; in turn, this conclusion amounts to Picard's theorem that a non-constant entire function attains all complex values but at most one: indeed, a solution would yield the non-constant entire function $x(z)$ never attaining the values $0, 1$ for $z \in \mathbb{C}$.

More generally, a holomorphic analogue of Siegel's theorem is the assertion that any affine smooth complex algebraic curve yields a hyperbolic Riemann surface[11] unless its genus is 0 and there are at most two points at infinity. Vojta brought this kind of analogy to a very far-reaching extent, and formulated important conjectures in higher dimensions, both for integral points in the classical sense and for the holomorphic context (see for instance [11] or [38] for an account).

[10] The results that we have mentioned so far predict degeneracy of integral (or rational) points. Conversely, when Vojta's above alluded conditions are not verified, one expects sometimes a dense set of rational points. The study of such situations involves rather different techniques; see [38] for an account up to 1991.

[11] By this we mean that there are no non-constant holomorphic maps from $\mathbb{C}$ to it. This notion well compares to other ones; see [38].

This provides still further evidence of the influence of Siegel's theorem in contemporary mathematics.

We could go further on with this list, *e.g.* by mentioning that Siegel's theorem has been proved in the context of Drinfeld modules by D. Ghioca and T.J. Tucker in [34], but since our aim here is not to give a complete account but just to give a selection of developments, we leave it with the remarks above.

Before going ahead and presenting (a sketch of) the said proofs, let us pause a further moment to introduce some notation and to recall Roth's theorem, which shall be used in two of the arguments.

3 Siegel's Theorem and some preliminaries

Let k be a number field and let M_k be the set of places of k.[12] For $v \in M_k$ we let k_v denote the completion of k at v. We normalize the absolute values $|\cdot|_v$ associated to $v \in M_k$ so that the product formula $\prod_v |x|_v = 1$ holds for $x \in k^*$ and so that the Weil absolute logarithmic height of $x \in k^*$ is $h(x) = \sum_v \log^+ |x|_v$, where $\log^+ t := \max(0, \log t)$ for t a positive real; we shall also write $H(x) = \exp h(x)$.

Usually, S will denote a finite set of places of k containing all the archimedean ones; in such cases we let $\mathcal{O}_S = \mathcal{O}_{k,S} = \{x \in k : |x|_v \le 1 \ \forall v \notin S\}$ (resp. $\mathcal{O}_S^* = \{x \in k : |x|_v = 1 \ \forall v \notin S\}$) denote the ring (resp. group) of S-integers (resp. S-units) in k: *i.e.*, their denominators (resp. and numerators) are divisible only by primes in S. (See [11] or [37] for the theory.)

With these conventions we state Lang's form of Roth's Theorem (appearing in [37]) as:

Theorem 3.1 (Roth's Theorem – general form). *Let k be a number field and let S be a finite subset of M_k. For $v \in S$, let $\alpha_v \in k_v$ be algebraic over k. Then, for every $\epsilon > 0$ and all $\beta \in k$ with the exception of a finite set depending on ϵ we have*

$$\prod_{v \in S} \min(1, |\beta - \alpha_v|_v) \ge H(\beta)^{-2-\epsilon}.$$

The statement even holds for $\alpha_v = \infty \in \mathbb{P}_1(k_v)$, on using the convention $|\beta - \infty| := 1/|\beta|$.

We note that the original form of Roth's theorem is recovered on taking $k = \mathbb{Q}$ and $S = \{\text{usual absolute value}\}$.

We shall prove Siegel's theorem in the following form:

Theorem 3.2. *Let C be an affine irreducible curve over a number field k, and suppose it has infinitely many integral points. Then C has genus 0 and at most two points at infinity.*

[12] For instance, when $k = \mathbb{Q}$ we have the place corresponding to the usual absolute value and the places corresponding to p-adic absolute values.

For completeness, let us recall first how to derive the case of genus 0 from Roth's Theorem 3.1. We can suppose that the curve is $\mathbb{P}_1 \setminus \{0, 1, \infty\}$; this may be embedded in $\mathbb{A}^2$ as the curve $x(1-x)z = 1$, so that the S-integral points correspond to solutions to the equation $x + y = 1$ in S-units x, y.

If $(x, y) = (x, 1 - x)$ is a solution, namely $x, 1 - x$ are both S-units, we set $\beta = x$ in Theorem 3.1. Also, we partition the places v in S into three disjoint sets S_1, S_2, S_3 according as $|x|_v \leq 1/2$, $1/2 < |x|_v \leq 2$ and $|x|_v > 2$ respectively. This partition of S may depend on the solution, but since S is finite we may suppose for our purposes that the partition is the same for all solutions. In the three cases we set resp. $\alpha_v = 0, 1, \infty$. We have $\min(1, |\beta - \alpha_v|_v) \leq |x|_v$ for $v \in S_1$, $\leq |y|_v$ for $v \in S_2$, and $\leq 1/|x|_v$ for $v \in S_3$.

On the other hand we have:

- $H(x) = H(x^{-1}) = \prod_{v\in S} \max(1, |x|_v^{-1}) \leq (\prod_{v\in S_1} |x|_v^{-1}) \cdot 2^{\#S}$;
- $H(y) = H(y^{-1}) = \prod_{v\in S} \max(1, |y|_v^{-1}) \leq (\prod_{v\in S_2} |y|_v^{-1}) \cdot 2^{\#S}$;
- $H(x) = \prod_{v\in S} \max(1, |x|_v) \leq (\prod_{v\in S_3} |x|_v) \cdot 2^{\#S}$,

where we have essentially used that $x, y \in \mathcal{O}_S^*$ to simplify the expression for the height.

Taking into account these cases we find, using also $H(y) = H(1 - x) \geq H(x)/2$,

$$\prod_{v\in S} \min(1, |\beta - \alpha_v|_v) \leq 8^{\#S} H(x)^{-2} H(1-x)^{-1} \leq 2^{3\#S+1} H(\beta)^{-3}.$$

Hence for large $H(x) = H(\beta)$ we get a contradiction with the inequality of Theorem 3.1 (with any $\epsilon < 1$).[13] This shows that $x = \beta$ belongs to a finite set, which concludes the argument.

The general case of positive genus seems not to fall into this pattern. We shall give three different arguments working in general. We shall be brief in our exposition, the purpose being merely to illustrate the main principles. Also, we shall rely on facts (like the behaviour of heights under rational maps) that Siegel had to develop by himself at that time.

4 Three arguments for Siegel's Theorem

4.1 A first argument for Siegel's Theorem

This argument follows lines which are the ones most commonly presented (see *e.g.* [37] or [46]).

[13] The exponent '-3' attributed to $H(x)$ corresponds to the 'three' points at infinity, *i.e.* $0, 1, \infty$, for the present curve. This leaves additional 'space' for Theorem 3.1 to be applied, and in fact we can allow x, y to be 'almost' S-units to obtain the same finiteness conclusion.

We assume that the curve C has genus $g > 0$, is embedded in some affine space $\mathbb{A}^n$ and that its projective closure $\tilde{C} \subset \mathbb{P}_n$ is non-singular (one may easily reduce the proof to this case).

Let us suppose now given an infinite sequence $\{P_i\}$ of distinct S-integral points on C.

By compactness of $\tilde{C}(k_v)$, on going to a still infinite subsequence we may suppose that for all $v \in S$ the sequence converges in the v-adic topology, say that $P_i \to_v Q_v$ for $i \to \infty$ and $v \in S$, where Q_v is some point in $\tilde{C}(k_v)$.

For a point $P \in \mathbb{A}^n(k_v)$ we let $|P|_v$ be the v-adic sup-norm of the coordinates. Since S is finite, we may assume, going again to an infinite subsequence, that there is a $v_0 \in S$ such that $|P_i|_{v_0} = \max_{v \in S} |P_i|_v$, for all points P_i.

We have $H(P_i) = \prod_{v \in M_k} \max(1, |P_i|_v) = \prod_{v \in S} \max(1, |P_i|_v)$[14], the last equality following from the fact that the P_i are S-integral points in k (which means $|P_i|_v \leq 1$ for $v \notin S$). Hence

$$H(P_i) \leq \max(1, |P_i|_{v_0})^{\#S}. \tag{4.1}$$

In particular, this says that $|P_i|_{v_0} \to \infty$ since the P_i are distinct, so Q_{v_0} is a point at infinity, namely in $\tilde{C} \backslash C$. Since $P_i \to_{v_0} Q_{v_0}$ we also have $d(P_i, Q_{v_0}) \ll |P_i|_{v_0}^{-\delta}$, where $d(\cdot\,,\cdot)$ is a distance function around $Q_{v_0} \times Q_{v_0}$ (associated for instance to some local parameter at Q_{v_0}), and δ is a positive number depending only on the geometry of $\tilde{C}$ around Q_{v_0} (namely on the choice of distance function). Hence, by (4.1),

$$d(P_i, Q_{v_0}) \ll H(P_i)^{-\delta/\#S}. \tag{4.2}$$

Now, the coordinates of Q_{v_0} are algebraic (because Q_{v_0} is a point in $\tilde{C} \setminus C$) and we may suppose they lie in k. Hence by Roth's Theorem 3.1 (applied with any subset of S containing $\{v_0\}$) we have

$$d(P_i, Q_{v_0}) \gg_\epsilon H(P_i)^{-2-\epsilon}.$$

If $\delta > 2\#S$ this is sufficient to contradict (4.2) and conclude the proof. This may happen, but only for special equations, *e.g.* of 'Thue's type', provided also $\#S = 1$.

To overcome this obstacle in general, Siegel embedded the curve in its Jacobian J (recall that $\tilde{C}$ has positive genus). By the weak Mordell–Weil Theorem for J one may write, for any fixed positive integer m, $P_i = mP_i' + R_i$ where the R_i are taken from a finite set and where the R_i, P_i' are points in $J(k)$. (In this argument the Chevalley–Weil Theorem suffices in place of Mordell–Weil - see [11] for these results. This may be important for effectiveness, not known for the Mordell–Weil Theorem. See especially the letter of Coates in the Appendix to [59] for a related viewpoint.)

[14] We use the affine height in this argument, contrary to the sequel.

Also, by going to a suitable infinite subsequence we may assume that $R_i = R$ is fixed and that also the P_i' converge with respect to any $v \in S$. (Note that the P_i' lie on the unramified cover of the translated curve $C - R$ obtained by inverse image under the multiplication-by-m map on J.)

If v_0 is as above, the P_i' will converge v_0-adically to a point Q' such that $mQ' + R = Q_{v_0}$, so Q' will be an algebraic point of J.

The advantage of these considerations comes from the different behaviour of the distance and of the height with respect to the multiplication and translation maps on J. We may choose a height (still denoted $H(\cdot)$) on J, associated to a symmetric very ample divisor (see [11], [46]), and choose a related embedding for J in some $\mathbb{P}_n$, which in turn induces a projective embedding for C. Now we have two important facts:

(i) Since both $P \mapsto mP$ and $P \mapsto P+R$ are unramified, we have $d(P_i', Q') \asymp d(P_i, Q_{v_0})$.
(ii) It is known ([11,46]) that we have $H(mP) \gg H(P)^{m^2(1-\epsilon)}$ and $H(P + R) \asymp H(P)$ whence, choosing *e.g.* $\epsilon = 1/2$, $H(P_i') \ll H(P_i)^{2/m^2}$.[15]

At this point, by (i) and Roth's Theorem 3.1 applied to the P_i', Q' we find $d(P_i, Q_{v_0}) \gg d(P_i', Q') \gg H(P_i')^{-2-\epsilon} \gg H(P_i')^{-3}$, say.

By (ii) we thus find $d(P_i, Q_{v_0}) \gg H(P_i)^{-6/m^2}$. Comparison with (4.2) yields finally boundedness of $H(P_i)$, *i.e.* finiteness, if $\delta > 6\#S/m^2$, which can be achieved by choosing m large enough. It will be noted that the full force of Roth's Theorem is not needed in this argument; however we need an exponent (in place of 2) which is fixed, *i.e.* not depending on the involved points and field of definition.

4.2 A second argument for Siegel's Theorem

We shall preserve the above notation. We also put $\tilde{C} \setminus C = \{A_1, \ldots, A_r\}$, where A_i are distinct points defined over k. We let $\{P_i\}$ be an infinite sequence of distinct S-integral points in $C(k)$, such that for each $v \in S$ we have $P_i \to_v Q_v$. (By compactness of $C(k_v)$, this assumption of convergence causes no loss of generality in proving finiteness.)

We start by observing that the general theorem may be reduced to the case $r \geq 3$, by the following principle: if the curve $\tilde{C}$ has positive genus, its fundamental group has $\mathbb{Z}$ as a quotient, and thus it admits unramified covers $\pi : \tilde{C}' \to \tilde{C}$ of arbitrary (finite) degree; moreover, (*e.g.* by specialization) we may assume that such cover is defined over $\overline{\mathbb{Q}}$. Note that if C is affine,

[15] This behaviour of the distance and height is similar to the simpler situation of the multiplicative group $\mathbb{G}_m$ where we have $d(x^m, 1) \asymp d(x, 1)$ for x near 1 and $H(x^m) = H(x)^{|m|}$.

$\pi^{-1}(C)$ is also affine, with $\geq \deg \pi$ points at infinity. Now, the Chevalley–Weil Theorem ([11,37]) implies that any sequence of (k, S)-integral points on C lifts to a sequence of (k', S')-integral points on $\pi^{-1}(C)$, for a suitable number field k' and finite set of places S' and then it suffices to apply the special case to the cover, provided its degree is ≥ 3.

In short, for Siegel's Theorem the crucial case occurs when $\#(\tilde{C} \setminus C) = r \geq 3$, as we shall suppose in the sequel.

We shall try to obtain a contradiction by means of Roth's Theorem 3.1 by looking at values $\varphi(P_i)$ of a non-constant rational function $\varphi \in k(C)$ on the sequence. Since $P_i \to_v Q_v$ we have $\varphi(P_i) \to_v \varphi(Q_v)$ for each $v \in S$. Therefore $\varphi(P_i) - \varphi(Q_v)$ will be v-adically small. To exploit Roth's Theorem 3.1 we shall therefore put $\beta = \beta_i := \varphi(P_i)$ and $\alpha_v = \varphi(Q_v)$ for a certain subset of the places v in S. Note that *a priori* $k(Q_v)$ may be a transcendental extension of k; in such cases this choice of α_v would not be legitimate. However we know that Q_v is algebraic (and we may even assume it is defined over k) provided $Q_v \in \tilde{C} \setminus C$ is a point at infinity; and on the other hand the places such that this does not happen give rise to bounded $|P_i|_v$, which is harmless. Therefore we put $S_j := \{v \in S \ : \ Q_v = A_j\}$ and define $\alpha_v = \varphi(Q_v)$ for v in $\cup_{j=1}^{r} S_j$ (disjoint union!).

We have now to estimate the various quantities occurring in Roth's Theorem 3.1.

Let $t_j \in k(C)$ be a local parameter at A_j; also, for a large fixed integer N consider the linear space $L(NA_j)$ of functions in $k(C)$ whose pole divisor is $\geq -NA_j$. Embedding $\tilde{C}$ in a projective space by means of $L(NA_j)$ produces (by definition) a Weil height which is $\asymp H_{A_j}^N$, where H_P is a height associated to a point P, up to a $O(1)$ factor (see [11]).

In calculating the height of P_i by means of this linear system, practically only the places in S_j contribute. In fact, if $g_1, \ldots, g_d$ is a basis for $L(NA_j)$, we have,

$$\begin{aligned} H_{NA_j}(P_i) &= H(g_1(P_i) : \ldots : g_N(P_i)) \\ &= \prod_v \sup_l |g_l(P_i)|_v \asymp \prod_{v \in S_j} \sup_l |g_l(P_i)|_v. \end{aligned}$$

This is because the P_i are S-integral points and the g_l are regular on $\tilde{C} \setminus \{A_j\}$, so $\sup_l |g_l(P_i)|_v$ is bounded for all $v \notin S_j$ and is 1 for almost all places. In turn, this yields

$$H_{A_j}^N(P_i) = H_{NA_j}(P_i) = H(g_1(P_i) : \ldots : g_N(P_i)) \ll \prod_{v \in S_j} |t_j(P_i)|_v^{-N},$$

namely

$$\prod_{v \in S_j} |t_j(P_i)|_v \ll H_{A_j}^{-1}(P_i). \tag{4.3}$$

Now, define $m_j \geq 1$ as the multiplicity of A_j as a zero of $\varphi - \varphi(A_j)$. (If φ has a pole of A_j we use the previous convention $\varphi - \varphi(A_j) = 1/\varphi$ so m_j will be the order as a pole.)

We have that $|\varphi(P_i) - \varphi(A_j)|_v \ll |t_j(P_i)|_v^{m_j}$ if $v \in S_j$. Hence

$$\prod_{j=1}^{r} \prod_{v \in S_j} \min(1, |\varphi(P_i) - \varphi(A_j)|_v) \ll \prod_{j=1}^{r} \prod_{v \in S_j} |t_j(P_i)|_v^{m_j} \ll \prod_{j=1}^{r} H_{A_j}^{-m_j}(P_i),$$

where the last inequality follows from (4.3). At this point, Roth's Theorem 3.1 (with $\cup_{j=1}^{r} S_j$ in place of S) bounds from below the left-hand side of the inequality for all but finitely many P_i, giving

$$H(\varphi(P_i))^{-2-\epsilon} \ll \prod_{j=1}^{r} H_{A_j}^{-m_j}(P_i),$$

where the height on the left is the usual Weil height of an algebraic number. For notational convenience we express this in terms of logarithmic heights $h(\cdot)$, obtaining

$$\sum_{j=1}^{r} m_j h_{A_j}(P_i) \leq (2 + o(1))h(\varphi(P_i)) + O(1), \qquad i \to \infty.$$

Now, we invoke the elementary theory of heights on curves (see [11,37] or [46]):

(i) All heights relative to a point are known to be asymptotically equivalent (see [11] or [46]), so, if we choose once and for all a point $B \in C(\bar{k})$ we have $h_{A_j}(P_i) = h_B(P_i)(1 + o(1)) + O(1)$ as $i \to \infty$.

(ii) The height $h \circ \varphi$ is equivalent (up to $O(1)$) to the height relative to the divisor of poles of φ, which in turn is (by (i)) equivalent to $\deg(\varphi) h_B$.[16]

Summing up we obtain

$$\sum_{j=1}^{r} m_j h_B(P_i)(1 + o(1)) \leq (2 + o(1)) \deg(\varphi) h_B(P_i) + O(1),$$

whence, dividing by $h_B(P_i)$ (which tends to ∞) and letting $i \to \infty$,

$$\sum_{j=1}^{r} \mathrm{ord}_{A_j}(\varphi - \varphi(A_j)) \leq 2 \deg(\varphi). \tag{4.4}$$

Hence, we can deduce a contradiction if we can find a non-constant function $\varphi \in k(C)$ such that (4.4) is *not* verified; in such cases the above argument

[16] It is remarkable that these facts in essence go back to the same paper of Siegel translated here.

provides a proof of the finiteness of integral points which essentially depends only on the Roth's Theorem 3.1.

A suitable function φ exists for instance in the case of Thue's curves; however this does not appear to be the case for a general curve. (See the discussion in [59].)

However, if we are allowed to use some geometric properties of abelian varieties, we can see how to contradict (4.4), and hence prove Siegel's Theorem, by replacing the curve $\tilde{C}$ with a cover of it and seeking the function φ on this new curve. This will provide a second general finiteness proof for S-integral points.[17]

We note that (4.4) implies $r \leq 2\deg(\varphi)$ (no matter the points A_i) and it is this last inequality that we shall eventually contradict. For this we have to find a rational function on C of 'small' degree. Let us note at once that for our purposes we may increase arbitrarily the number field k, so it will suffice to work with the function field of the curve over $\overline{\mathbb{Q}}$.

Let us then denote by $\gamma(C)$ the so-called *gonality* of C, that is the minimum degree of a non-constant rational function on C (defined over the algebraic numbers).[18] Using the Riemann-Roch theorem it is an easy matter to see that $\gamma(C) \leq g+1$ and one may improve this to $\gamma(C) \leq (g+3)/2$ (see for example the discussion at pp. 154-159 of [2]). However such inequalities are plainly not sufficient for our aim, since r may be quite small.

With the purpose of increasing r, while keeping $\deg(\varphi)$ not too large, we shall then go to a cover of the curve; we have already done this at the end of the first proof, to achieve the condition $r \geq 3$ needed therein. For the present proof we require a much more substantial gain. To obtain a suitable cover, we embed the curve in its Jacobian J and define $\tilde{C}_n$ as the inverse image of $\tilde{C}$ under the multiplication-by-n map $[n]$; it is well known that $\tilde{C}_n$ is an irreducible unramified cover of $\tilde{C}$, of degree n^{2g}. Now we have a fundamental property, of interest in itself, which appears in Siegel's paper translated here:

Lemma 4.1. *With the above notation, for large n we have $\gamma(\tilde{C}_n) \leq gn^{2g-2}\gamma(\tilde{C})$.*

Note that the inequality $\gamma(\tilde{C}_n) \leq n^{2g}\gamma(\tilde{C})$ is obvious, but would not be sufficient here, the gain of a factor tending to zero as $n \to \infty$ being vital (here we gain gn^{-2}). This lemma was found by Siegel and also used in his proof.[19] Note

[17] We remark that this last approach has substantial similarities with a proof given by Robinson and Roquette (who also used Roth's Theorem), formulated in the realm of non-standard arithmetic. See their paper [44], where some account of 'non-standard' prerequisites also appears. For instance they derive inequalities which remind of (4.4) and follow from it; however they use the concept of 'exceptional divisors', which here is totally absent and in fact has no use in this context.

[18] This last condition is immaterial if C is defined over $\overline{\mathbb{Q}}$.

[19] Siegel's statement was in slightly different form and for a somewhat different purpose. For Siegel's viewpoint see also the letter by Coates in the Appendix to [59].

that we have not needed it in our presentation of the first argument; we could avoid it because we used heights on J, whose behaviour under the map $[n]$ boils down to the same principles which imply the lemma.

Siegel proved the lemma by computing the degree of a suitable function as the number of its zeros, and by using at this stage the principle of variation of the angle (applied to the logarithm of suitable theta functions). Instead we shall now sketch a geometric formulation of Siegel's proof, using some intersection theory and known formulas from the theory of abelian varieties (see [11,46]).[20]

Sketch of proof of Lemma 4.1. We follow Siegel's idea. Let φ be a non-constant rational function on $\tilde{C}$ and, for $P_1, \ldots, P_g$ points on $\tilde{C}$ where φ is defined, set $\phi(P_1, \ldots, P_g) = \varphi(P_1) \cdots \varphi(P_g)$. This is a symmetric function of the g points P_i. We view $\tilde{C}$ embedded in J, which is birationally equivalent to the g-th symmetric power of $\tilde{C}$. Through this embedding ϕ induces a rational function on J. For large n one can prove that $\tilde{C}_n$ is generically in the domain of definition of ϕ (note that $\tilde{C}_n$ 'tends' to become dense everywhere on J); so by restriction we obtain a rational function on $\tilde{C}_n$, which we denote again by ϕ. (Note that in general this will be different from the natural function $\varphi \circ [n]$ induced by φ on $\tilde{C}_n$, which would not produce anything useful.)

We shall bound $\deg(\phi)$, as a function on $\tilde{C}_n$; for this we bound the number of zeros. It follows easily from the very definition that each zero Q of φ on $\tilde{C}$ produces a zero-set of ϕ on J which is $Q + P_2 + \ldots + P_g$, where the P_i run along $\tilde{\mathbb{C}}$ and the sum refers to the group operation in J. In usual notation, this is $Q + \Theta$, where Θ is called the 'Theta divisor' on J; we may also suppose that the embedding of C in J is chosen so that Θ is a symmetric divisor.

The zero Q of φ on $\tilde{C}$ will then produce $\leq (Q + \Theta).\tilde{C}_n$ zeros of ϕ on $\tilde{C}_n$, where (.) denotes intersection product. Since translation does not change the algebraic equivalence class of a divisor, we also have $\deg(\phi) \leq \deg(\varphi)(\Theta.\tilde{\mathbb{C}}_n) = \deg(\varphi)(\Theta.[n]^*\tilde{C})$.

To compute the intersection product on the right note that

$$([n]^*\Theta.[n]^*\tilde{C}) = n^{2g}(\Theta.\tilde{C}) = n^{2g}g,$$

by the known formula $(\Theta.\tilde{C}) = g$ and since $\deg[n] = n^{2g}$. On the other hand the class of $[n]^*\Theta$ in $\mathrm{Pic}(J)$ is known to be n^2 times the class of Θ, hence $([n]^*\Theta.[n]^*\tilde{C}) = n^2(\Theta.[n]^*\tilde{C})$.

Summing up, we have found $\deg(\phi) \leq \deg(\varphi)gn^{2g-2}$ for large n; choosing φ as a non-constant function on $\tilde{C}$ of minimal degree finishes the proof of the lemma. □

We can now complete the second proof of Siegel's Theorem. Suppose that C has infinitely many S-integral points over k. Then, by the Chevalley–Weil

[20]The paper [44] takes this lemma for granted and refers to Siegel's proof and to another proof by Roquette, at that time unpublished.

Theorem (as at the end of the second argument) for each positive integer n the curve C_n, *i.e.* the inverse image of C in $\tilde{C}_n$, would have infinitely many S_n-integral points, over a suitable finite extension k_n of k, S_n being a suitable finite set of places of k_n. Let φ_n be a non-constant function on C_n, of degree $\gamma(C_n)$, where we may suppose it is defined on k_n. Note that C_n has rn^{2g} points at infinity. Then (4.4), applied with C_n in place of C, φ_n in place of φ, gives $rn^{2g} \le 2\deg(\varphi_n) = 2\gamma(C_n)$. However this inequality is inconsistent with the lemma if n is large enough. This contradiction completes the argument.

For still another viewpoint and approach (especially related to this last one) see the letter of Coates to Cassels, in the Appendix to [59].

4.3 A third argument for Siegel's Theorem

Beyond Roth's Theorem, the first of the arguments outlined above uses results on Jacobians and their arithmetic which, though nowadays 'routine', remain somewhat delicate. In view also of the fact that Roth's Theorem suffices for special equations, it is therefore sensible to ask for even more direct deductions of Siegel's Theorem. The second argument above still uses Jacobians (though in a less heavy way than above). We shall now illustrate a direct argument avoiding Jacobians and using essentially only Diophantine approximation, valid for affine curves with at least three points at infinity; this was given by Corvaja and the second author in [13], and the method leads also to some results in higher dimensions (mentioned above).

This proof, which we shall soon reproduce, uses an embedding of the curve into projective spaces of high dimension (in order to reproduce some features of Thue curves, which however do not occur generally). For this reason, it requires not just Roth's Theorem, but a higher-dimensional extension of it. This deep and far-reaching extension was found originally by W.M. Schmidt around 1970. As for Roth's Theorem, the statements evolved to include eventually arbitrary number fields and finite sets of places. In the sequel we shall refer to a version by H.P. Schlickewei; as above we denote by S a finite set of places of k containing the archimedean ones and by $|\mathbf{x}|_v$ the v-adic sup norm, for $\mathbf{x} \in k^n$; we also denote by $H(\mathbf{x}) := \prod_v |\mathbf{x}|_v$ the *projective* Weil height of the vector $\mathbf{x}$.

Theorem 4.2 (Schmidt's Subspace Theorem). *For $v \in S$ let $L_{iv}, i = 1, \dots, d$, be linearly independent linear forms with coefficients in $k_v \cap \overline{\mathbb{Q}}$ and let $\epsilon > 0$. Then for all $\mathbf{x} \in k^n$ with the exception of a finite union (depending on ϵ) of hyperplanes of k^n, we have*

$$\prod_{v \in S} \prod_{i=1}^{d} \frac{|L_{iv}(\mathbf{x})|_v}{|\mathbf{x}|_v} \ge H(\mathbf{x})^{-d-\epsilon}.$$

See *e.g.* [11] for a proof. It is easy to recover Roth's Theorem 3.1 from the case $d = 2$ (projective dimension 1) of this statement, on setting $\mathbf{x} = (x_1, x_2) = (\beta, 1)$, $L_{1v}(\mathbf{x}) = x_2$, $L_{2v}(\mathbf{x}) = x_1 - \alpha_v x_2$.

Now we shall see, following [13], how this leads to a quick proof of Siegel's Theorem.

We have already remarked that we may work under the assumption that $r := \#(\tilde{C} \setminus C) \geq 3$.

We denote by $A_1, \ldots, A_r$ the points at infinity and we suppose on enlarging k that they are defined over k. The main principle now will be to embed C in a projective space of sufficiently large dimension (instead of the Jacobian) so that $\tilde{C}$ has 'large order contact' with suitable hyperplanes, at the points A_i at infinity. This will reproduce, in higher dimension, the favourable situation which occurs with the plane curves of Thue-type.

For a large but fixed integer N, we consider the vector space

$$V = V_N := \{f \in k(C) \ : \ \mathrm{div}(f) \geq -N(A_1 + \ldots + A_r)\}$$

of rational functions on C (over k) with poles only at infinity, of orders at most N. By (a weak form of) Riemann-Roch we have $d = d_N := \dim V_N \geq rN + O(1)$. Let $f_1, \ldots, f_d$ be a basis of V.

If, as above, $\{P_i\}$ is an infinite sequence of S-integral points, we may assume that they converge v-adically for each $v \in S$ to points $Q_v \in C(k_v)$. We let S' be the set of places such that Q_v is one of the A_i and we let $S'' = S \setminus S'$.

By multiplying the f_j by suitable non-zero constants we shall have $f_j(P_i) \in \mathcal{O}_S$ for all i, j.

Observe that for $v \in S''$ the values $|f_j(P_i)|_v$ are uniformly bounded, since $P_i \to_v Q_v \in C(k_v)$ and since the f_j are regular on C.

For fixed $v \in S'$ we consider the filtration of $V \supset W_1 \supset W_2 \supset \ldots$ defined by the spaces

$$W_j = W_{j,v} := \{f \in V \ : \ \mathrm{ord}_{Q_v} f \geq j - 1 - N\}.$$

We have $\dim(W_j / W_{j+1}) \leq 1$, since increasing by 1 the order of zero at a point imposes at most one linear condition, so in particular $\dim W_j \geq d - j + 1$.

Now we may pick a basis for W_d and complete it successively to bases for $W_{d-1}, W_{d-2}, \ldots, W_1$, obtaining vectors $w_d, w_{d-1}, \ldots, w_1$. Expressing these vectors in terms of $f_1, \ldots, f_d$ we obtain linear forms $L_{dv}, \ldots, L_{1v}$ in the f_j, defined over k (since now $v \in S'$) and such that

$$\mathrm{ord}_{Q_v} L_{jv} \geq j - 1 - N, \qquad j = 1, \ldots, d.$$

For $v \in S''$ we define $L_{jv} := f_j$.

For $v \in S'$ let $t_v \in k(C)$ be a local parameter at Q_v. Then the last displayed formula yields

$$|L_{jv}(P_i)|_v \ll |t_v(P_i)|_v^{j-1-N}, \qquad j = 1, \ldots, d.$$

On the other hand $|L_{jv}(P_i)|_v \ll 1$ for $v \in S''$ (since the $f_j(P_i)$ are then bounded). Hence

$$\prod_{v\in S}\prod_{j=1}^{d}|L_{jv}(P_i)|_v \ll \Big(\prod_{v\in S'}|t_v(P_i)|_v\Big)^{d(d-2N-1)/2}. \tag{4.5}$$

Put now $\mathbf{x}=\mathbf{x}_i := (f_1(P_i),\ldots,f_d(P_i))$. Since the values $f_j(P_i)$ are S-integers, we have $|\mathbf{x}|_v \le 1$ for $v \notin S$, so $H(\mathbf{x}) = \prod_{v\in S}|\mathbf{x}|_v$. Hence $\prod_{v\in S}\prod_{j=1}^{d}|\mathbf{x}|_v = H(\mathbf{x})^d$.

Moreover, as already noted, we have $|\mathbf{x}|_v \ll 1$ for $v \in S''$, since the $f_j(P_i)$ are then v-adically bounded; for $v \in S'$ the f_i have a pole of order at most N at Q_v, so $|\mathbf{x}|_v \ll |t_v(P_i)|_v^{-N}$. This yields $H(\mathbf{x}) \ll (\prod_{v\in S'}|t_v(P_i)|_v)^{-N}$, *i.e.* $\prod_{v\in S'}|t_v(P_i)|_v \ll H(\mathbf{x})^{-1/N}$.

Finally, writing $L_{jv}(P_i) = L_{jv}(\mathbf{x})$ as linear forms evaluated at $\mathbf{x}_i$, we get by (4.5)

$$\prod_{v\in S}\prod_{j=1}^{d}\frac{|L_{jv}(\mathbf{x})|_v}{|\mathbf{x}|_v} \ll \Big(\prod_{v\in S'}|t_v(P_i)|_v\Big)^{d(d-2N-1)/2} H(\mathbf{x})^{-d} \ll H(\mathbf{x})^{-d-\eta}$$

where $\eta = d(d-2N-1)/2N$ is > 0 if N has been chosen large enough (recall $d \ge 3N + O(1)$).

Application of Schmidt's Subspace Theorem 4.2 then implies that all the vectors $\mathbf{x}_i$ lie in a finite union of hyperplanes. However, lying in a given hyperplane gives an equation

$$\sum_{j=1}^{d} c_j f_j(P_i) = 0$$

with fixed coefficients c_j not all zero; but then the function $\sum_{j=1}^{d} c_j f_j$ is nonzero and may have only finitely many zeros P_i. This concludes the argument.

In the argument the relevance of the condition '$r \ge 3$' appears clearly (and in fact this condition cannot be omitted if the curve has genus 0).

References

[1] Y. André, "*G*-Functions and Geometry", Aspects of Mathematics, E13, Friedr. Vieweg & Sohn, Braunschweig, 1989.

[2] E. Arbarello, M. Cornalba, P. A. Griffiths and J. Harris, "Geometry of Algebraic Curves", Springer-Verlag (GdmW **267**), 1985.

[3] P. Autissier, *Points entiers sur les surfaces arithmétiques*, J. Reine Angew. Math. **531** (2001), 201–235.

[4] A. Baker and G. Wüstholz, "Logarithmic Forms and Diophantine Geometry", Vol. I, Cambridge Univ. Press (NMM **9**), 2007.

[5] A. BÉRCZES, K. GYŐRY and J.-H. EVERTSE, *Effective results for unit equations over finitely generated integral domains*, Math. Proc. Cambridge Philos. Soc. **154** (2013), 351–380.

[6] Y. BILU, *Effective analysis of integral points on algebraic curves*, Israel J. Math. **90** (1995), 235–252.

[7] Y. BILU, *Quantitative Siegel's theorem for Galois coverings*, Compositio Math. **106** (1997), 125–158.

[8] E. BOMBIERI, "On G-Functions", Recent progress in analytic number theory, Vol. 2 (Durham, 1979), 1-67, Academic Press, London-New York, 1981.

[9] E. BOMBIERI, *The Mordell conjecture revisited*, Ann. Scuola Norm. Sup. Pisa Cl. Sci. (4) **17** (1990), 615–540; Errata-corrige: ibidem **18** (1991), 473.

[10] E. BOMBIERI, *Effective Diophantine approximation on* $\mathbb{G}_m$, Annali Scuola Normale Sup. Pisa Cl. Sci. **20** (1993), 61–89.

[11] E. BOMBIERI and W. GUBLER, "Heights in Diophantine Geometry", Cambridge Univ. Press (NMM **4**), 2006.

[12] E. BOMBIERI and W. M. SCHMIDT, *On Thue's equation*, Invent. Math. **88** (1987), 69–81; Correction: ibidem **97** (1989), 445.

[13] P. CORVAJA and U. ZANNIER, *A subspace theorem approach to integral points on curves*, C. R. Acad. Sci. Paris **334** (2002), 267–271.

[14] P. CORVAJA and U. ZANNIER, *On the number of integral points on algebraic curves*, J. Reine Angew. Math. **565** (2003), 27–42.

[15] P. CORVAJA and U. ZANNIER, *On a general Thue's equation*, Amer. J. Math. **126** (2004), 1033–1055; Addendum: ibidem **128** (2006), 1057–1066.

[16] P. CORVAJA and U. ZANNIER, *On integral points on surfaces*, Ann. of Math. (2) **160** (2004), 705–726.

[17] P. CORVAJA and U. ZANNIER, *On the integral points on certain surfaces*, Int. Math. Res. Not. 2006, Art. ID 98623.

[18] P. CORVAJA and U. ZANNIER, *Some cases of Vojta's conjecture on integral points over function fields*, J. Differential Geom. **17** (2008), 295–333; Addendum: Asian J. Math. **14** (2010), 581–584.

[19] P. CORVAJA, A. LEVIN and U. ZANNIER, *Integral points on threefolds and other varieties*, Tohoku Math. J. (2) **61** (2009), 589–601.

[20] J.-H. EVERTSE, "Upper Bounds for the Numbers of Solutions of Diphantine Equations", Mathematical Centre Tracts, Vol. 168, Mathematisch Centrum, Amsterdam, 1983.

[21] J.-H. EVERTSE, *An improvement of the quantitative subspace theorem*, Compositio Math. **101** (1996), 225–311.

[22] J.-H. EVERTSE and K. GYŐRY, *On the number of solutions of the weighted unit equation*, Compositio Math. **66** (1988), 329–354.

[23] J.-H. EVERTSE, K. GYŐRY, C. L. STEWART and R. TIJDEMAN, *On S-unit equations in two unknowns*, Invent. Math. **92** (1988), 461–477.

[24] J.-H. EVERTSE and B. EDIXHOVEN, "Diophantine Approximation and Abelian Varities", Introductory lecture. Papers from the conference held in Soesterberg, April 12-16, 1992, Springer-Verlag (LNM **1566**), 1993.

[25] J.-H. EVERTSE and R. FERRETTI, *A generalization of the Subspace Theorem with polynomials of higher degree*, Diophantine approximation, 175–198, Dev. Math., Vol. 16, Springer-Verlag, 2008.

[26] J.-H. EVERTSE and R. FERRETTI, *inequalities on projective varieties*, Int. Math. Res. Not. **25** (2002), 1295–1330.

[27] J.-H. EVERTSE and R. FERRETTI, *A further improvement of the quantitative subspace theorem*, Ann. of Math. (2) **177** (2013), 513–590.

[28] J.-H. EVERTSE and H. P. SCHLICKEWEI, *A quantitative version of the absolute subspace theorem*, J. Reine Angew. Math. **548** (2002), 21–127.

[29] G. FALTINGS, *Endlichkeitsätze für abelsche Varietäten über Zahlkörpern*, Invent. Math. **73** (1983), 349–366; Erratum: ibidem **75** (1984), 381.

[30] G. FALTINGS, *Diophantine approximation on abelian varieties*, Ann. of Math. (2) **133** (1991), 549–576.

[31] G. FALTINGS, *The general case of S. Lang's conjecture*, In: "Barsotti Symposium in Algebraic Geometry" (Abano Terme, 1991), Academic Press, San Diego (Perspect. Math. **15**), 1994.

[32] G. FALTINGS, *A new application of Diophantine approximations*, In: "A panorama of number theory or the view from Baker's garden" (Zürich, 1999), 231–246, Cambridge Univ. Press, 2002.

[33] G. FALTINGS, *Mathematics around Kim's new proof of Siegel's Theorem*, Diophantine Geometry, 173–188, CRM Series, 4, Ed. Norm., Pisa, 2007.

[34] D. GHIOCA and T. J. TUCKER, *Siegel's theorem for Drinfeld modules*, Math. Ann. **339** (2007), 37–60.

[35] M. HINDRY and J. H. SILVERMAN, "Diophantine Geometry", Springer-Verlag, GTM **201**, 2000.

[36] M. KIM, *The motivic fundamental group of* $\mathbb{P}^1\setminus\{0, 1, \infty\}$ *and the theorem of Siegel*, Invent. Math. **161** (2005), 629–656.

[37] S. LANG, "Fundamentals of Diophantine Geometry", Springer-Verlag, 1983.

[38] S. LANG, "Number Theory III", Encycl. of Mathematical Sciences, Vol. 60, Springer-Verlag, 1991.

[39] A. LEVIN, *Generalizations of Siegel's and Picard's theorems*, Ann. of Math. (2) **170** (2009), 609–655.

[40] A. LEVIN, *Siegel's theorem and the Shafarevich conjecture*, J. Théor. Nombres Bordeaux **24** (2012), 705–727.

[41] K. MAHLER, *A remark on Siegel's theorem on algebraic curves*, Mathematika **2** (1955), 116–127.

[42] D. POULAKIS, *Bound for the size of integral points on curves of genus zero*, Acta Math. Hungar. **93** (2001), 327–346.

[43] M. RAM MURTY and H. PASTEN, *Modular forms and effective Diophantine approximation*, J. Number Theory **133** (2013), 3739–3754.

[44] A. ROBINSON and P. ROQUETTE, *On the finiteness theorem of Siegel and Mahler concerning Diophantine Equations*, J. Number Theory **7** (1975), 121–176.

[45] K. F. ROTH, *Rational approximations to algebraic numbers*, Mathematika **2** (1955), 1–20.

[46] J-P. SERRE, "Lectures on the Mordell-Weil Theorem", Vieweg, 1990.

[47] C. L. SIEGEL, *The integer solutions of the equation* $y^2 = ax^n + bx^{n-1} + \cdots + k$, J. London Math. Soc. **1** (1926), 66–68.

[48] C. L. SIEGEL, *Über einige Anwendungen diophantischer Approximationen*, Abh. Preuß, Akad. Wissen. Phys.-math. Klasse, 1929 = Ges. Abh. Bd. I, Springer-Verlag 1966, 209–266.

[49] H. P. SCHLICKEWEI, *An explicit upper bound for the number of solutions of the S-unit equation*, J. Reine Angew. Math. **406** (1990), 109–120.

[50] H. P. SCHLICKEWEI, *S-unit equations over number fields*, Invent. Math. **102** (1990), 95–107.

[51] H. P. SCHLICKEWEI, *The quantititative subspace theorem for number fields*, Compositio Math. **82** (1992), 245–273.

[52] W. M. SCHMIDT, "Diophantine Approximation", Springer-Verlag (LNM **785**), 1980.

[53] W. M.SCHMIDT, "Dioiphantine Approximations and Diophantine Equations", Springer-Verlag (LNM **1467**), 1991.

[54] P. VOJTA, *Siegel's theorem in the compact case*, Ann. of Math. (2) **133** (1991), 509–548.

[55] U. ZANNIER, *Fields containing values of algebraic functions and related questions*, Number theory (Paris, 1993-1994), 199–213, London Math. Soc. Lecture Note Ser., 235, Cambridge Univ. Press, Cambridge, 1996.

[56] U. ZANNIER, *A brief history of the study of integral points on algebraic curves, particularly Siegel's theorem*, Geometry Seminars. 2001-2004 (Italian), 145–162, Univ. Stud. Bologna, 2004.

[57] U. ZANNIER, *On the integral points on certain algebraic varieties*, In: "European Congress of Mathematicians", Eur. Math. Soc., Zürich, 2005, 529–546.

[58] U. ZANNIER, *On the integral points on the complement of ramification-divisors*, J. Inst. Math. Jussieu **4** (2005), 317–330.

[59] U. ZANNIER, *Roth's theorem, Integral points and certain ramified Covers of* $\mathbb{P}_1$, In: "Analytic Number Theory", Cambridge Univ. Press, 2009, 471–491.

MONOGRAPHS

This series is a collection of monographs on advanced research topics of current interest in the fields of mathematics and physics.

Published volume

1. A. HOLEVO, *Probabilistic and Statistical Aspects of Quantum Theory*, 2011. ISBN 978-88-7642-375-8
2. U. ZANNIER (editor), *On Some Applications of Diophantine Approximations*, a translation of C. L. Siegel's *Über einige Anwendungen diophantischer Approximationen*, by C. Fuchs, with a commentary and the article by C. Fuchs and U. Zannier, 2014. SBN 978-88-7642-519-6, e-ISBN 978-88-7642-520-2

Volumes published earlier

QUADERNI

E. DE GIORGI, F. COLOMBINI, L. C. PICCININI, *Frontiere orientate di misura minima e questioni collegate*, 1972.

C. MIRANDA, *Su alcuni problemi di geometria differenziale in grande per gli ovaloidi*, 1973.

G. PRODI, A. AMBROSETTI, *Analisi non lineare*, 1973.

C. MIRANDA, *Problemi di esistenza in analisi funzionale*, 1975 (out of print).

I. T. TODOROV, M. MINTCHEV, V. B. PETKOVA, *Conformal Invariance in Quantum Field Theory*, 1978.

A. ANDREOTTI, M. NACINOVICH, *Analytic Convexity and the Principle of Phragmén-Lindelöf*, 1980.

S. CAMPANATO, *Sistemi ellittici in forma divergenza. Regolarità all'interno*, 1980.

Topics in Functional Analysis 1980-81, Contributors: F. STROCCHI, E. ZARANTONELLO, E. DE GIORGI, G. DAL MASO, L. MODICA, 1981.

G. LETTA, *Martingales et intégration stochastique*, 1984.

Old and New Problems in Fundamental Physics, in honour of GIAN CARLO WICK, 1986.

Interaction of Radiation with Matter, in honour of ADRIANO GOZZINI, 1987.

M. MÉTIVIER, *Stochastic Partial Differential Equations in Infinite Dimensional Spaces,* 1988.

Symmetry in Nature, in honour of LUIGI A. RADICATI DI BROZOLO, 2 voll., 1989 (reprint 2005).

Nonlinear Analysis, in honour of GIOVANNI PRODI, 1991 (out of print).

C. LAURENT-THIÉBAUT, J. LEITERER, *Andreotti-Grauert Theory on Real Hypersurfaces,* 1995.

J. ZABCZYK, *Chance and Decision. Stochastic Control in Discrete Time,* 1996 (reprint 2000).

I. EKELAND, *Exterior Differential Calculus and Applications to Economic Theory,* 1998 (2000).

Electrons and Photons in Solids, A Volume in honour of FRANCO BASSANI, 2001.

J. ZABCZYK, *Topics in Stochastic Processes,* 2004.

N. TOUZI, *Stochastic Control Problems, Viscosity Solutions and Application to Finance,* 2004.

CATTEDRA GALILEIANA

P. L. LIONS, *On Euler Equations and Statistical Physics,* 1998.

T. BJÖRK, *A Geometric View of the Term Structure of Interest Rates,* 2000 (2001).

F. DELBAEN, *Coherent Risk Measures,* 2002.

W. SCHACHERMAYER, *Portfolio Optimization in Incomplete Financial Markets,* 2004.

D. DUFFIE, *Credit Risk Modeling with Affine Processes,* 2004.

H. M. SONER, *Stochastic Optimal Control in Finance,* 2004 (2005).

LEZIONI LAGRANGE

C. VOISIN, *Variations of Hodge Structure of Calabi-Yau Threefolds,* 1998.

LEZIONI FERMIANE

R. THOM, *Modèles mathématiques de la morphogénèse,* 1971.

S. AGMON, *Spectral Properties of Schrödinger Operators and Scattering Theory,* 1975.

M. F. ATIYAH, *Geometry of Yang-Mills Fields,* 1979.

KAC M., *Integration in Function Spaces and some of its Applications,* 1983 (out of print).

J. MOSER, *Integrable Hamiltonian Systems and Spectral Theory,* 1983.

T. KATO, *Abstract Differential Equations and Nonlinear Mixed Problems,* 1988.

W. H. Fleming, *Controlled Markov Processes and Viscosity Solution of Nonlinear Evolution Equations*, 1988.

V. I. Arnold, *The Theory of Singularities and its Applications*, 1991.

J. P. Ostriker, *Development of Larger-Scale Structure in the Universe*, 1993.

S. P. Novikov, *Solitons and Geometry*, 1994.

L. A. Caffarelli, *The Obstacle Problem*, 1999.

Fotocomposizione "CompoMat" Loc. Braccone, 02040 Configni (RI) Italy
Finito di stampare nel mese di novembre 2014
dalla CSR srl, Via di Pietralata, 157, 00158 Roma